10 IDEAS TO MAKE SAFETY SUCK LESS

WRITTEN BY

SAM GOODMAN

A Pale Horse Media Co. Book

10 Ideas to Make Safety Suck Less

ISBN 9798843067748

Published by Pale Horse Media Co.

MORE BOOKS BY SAM GOODMAN

"Safety Sucks! The BULL $H!* in the Safety Profession They Don't Tell You About." First Edition. S. Goodman 2020

"Obscured: The Pursuit of Radical Self-acceptance." S. Goodman 2020

"Safety Sucks! The BULL $H!* in the Safety Profession They Don't Tell You About." The Expanded Second Edition. S. Goodman 2021

"In His Name." S. Goodman 2021

"Safety Sucks: The Manifesto." S. Goodman and I. Allison 2021

"WTFRM? A Reflection on What is Meaningful to Workplace Safety." S. Goodman 2021

"The Care and Feeding of Safety Practitioners: A brief guide to navigating the perils of the safety profession." S. Goodman 2022

DEDICATION

To every person out there fighting tirelessly to make things better. You are leading the way.

Sam Goodman – "The HOP Nerd"

CONTENTS

11

12

A QUICK CHAT BEFORE WE BEGIN

Welcome and thank you for picking up *10 Ideas to Make Safety Suck Less*! It truly means a lot to me that you would spend your hard-earned cash on a copy of this book. On that note, allow me to kick things off by saying that I hope you find value contained within the following pages, that this work inspires you to go out and seek to make things better, to make safety suck less, that you find practical and "doable" ideas contained within, and that when you place this book back on the shelf or gift it to a friend, that you feel that it was money well spent.

A little about me

I have weaved bits and pieces of my personal experience as a leader, HOP practitioner, nerd, and general member of "team human" throughout this book. Rather than vomiting my self-admiration all over these pages explaining to you who and what I am, providing you with some boastful list of my accomplishments, first allow me to expand upon what and who I am not. I am not a safety guru, god, prophet, or expert. I am not the "great knower of safety things." I do not hold all of the answers and I do not possess some miracle cure to all of your sources of pain around workplace safety. If that's what you were looking for when you picked up this book, my apologies, you will not find that here (or anywhere else for that matter).

At my core I am simply a human-centered safety practitioner, a betterment nerd, and overall believer that work shouldn't suck, even the stuff at work that relates to safety. I dedicate the vast majority of my time and energy into those things, I'm in constant pursuit of doing Safety Better. Even if that means stepping on some of our most sacred beliefs about safety, even if that means toppling some of our most sacred safety cows.

Because I refuse to submit to the current safety political landscape, because I refuse to cower in the shadow of "big safety" and refuse to be fearful of challenging our most hallowed beliefs, because of my disgust towards accepting things how they have always been, because of my refusal to tiptoe around sacred safety ideologies – many of which are horrific and harmful tools of blame and shame – because I regularly exclaim "bull shit" when I see bull shit, I get quite a bit of heat. Some have called me a provocateur, an ass hole, a disruptor, or flippant... All of which contain some bits of truth. But these are titles that I'll happily wear; they tell me that I'm on the right path. So, you might know me as an antihero, as an ass hole, or maybe a villain, but I'm really just a guy on a mission to make safety suck less.

At the time of publishing, I have spent the last 15-plus years working within the safety profession, with more than half of that time focused on leading organizations on the journey towards Human and Organizational Performance. I am a Human and Organizational Performance practitioner – meaning that I actually practice these concepts in real-life – starting first with several large organizations I worked for directly, and now mostly focusing on helping others on their HOP journeys. I state this not to add any undue or unproven credibility to this book, I would rather the ideas contained herein stand on their own two feet. I state this in an effort to express my understanding of the challenges faced when bringing about change in the real-world, in the corporate world, and to big and slow organizations that would often rather go bankrupt before even considering change. I have been there, I have stared that monster in the face, and I am here to tell you that it is very possible – although often challenging and frustrating – it is possible to bring about positive and sweeping change within your work world.

Arriving at these ideas

I found myself writing this book after years of Human and Organizational Performance (and safety) teaching and consulting. In addition to the *5 Principles of Human and Organizational Performance* brought to us by Todd Conklin, and the various other key teachings of folks like Bob

Edwards, Andrea Baker, Clive Lloyd, David Provan, Sidney Dekker, and many more – teachings that are sewn into the fabric of what we often refer to as *HOP, Safety Differently*, or *Safety II* – these ten ideas are some of the key pieces that I have found myself teaching almost daily for the last several years.

These 10 thoughts and ideas are the culmination of hundreds (if not thousands) of hours spent talking with practitioners, professionals, academics, researchers, and safety thinkers on *The HOP Nerd Podcast*. They are a product of my own views, thoughts, and opinions on what is meaningful to the safety of work. They are the result of years of my lived experiences (good and bad) as a safety professional and as a Human and Organizational Performance practitioner – trying things out, experimenting, and learning from those throughout high-risk industries and organizations.

These ideas and thoughts have been distilled down by my lived experiences. Experiences of working within traditional safety organizations and leading them through this shift towards doing safety better, by years of being a HOP practitioner, and by numerous hours spent teaching and consulting for organizations at various points along their Human and Organizational Performance journeys.

Why ten ideas?

Why not six? Twenty-three? Four? And on, and on… The answer is not mythical or magical, ten is simply where I found myself. With each passing year as a practitioner of Human and Organizational Performance, after being a part of multiple organizations journeys along the HOP path, after witnessing the good, the bad, and the ugly as it relates to creating progress in this space, these are some of the most impactful items that those seeking to do safety better must spend their days focusing on. These are the things that get us stuck, the not great things that we cling to, the amazing things that we avoid like the plague, and the pieces that – if we get them right – will propel us, build momentum, and create betterment along the way. These thoughts, ideas, and concepts are the shifts we must undertake as organizations as we move towards doing safety better, they are the things we must pump the brakes on and rapidly stop doing, the things that we must swiftly start doing, and the items we must insanely focus on doing better, if our hope is to craft environments in which work can go well.

I call these items "ideas" quite purposefully. So often we find ourselves as organizations and individuals seeking out quick and easy programs, step-by-step guides, and off-the-shelf products to cure the problems that ail us. Human and Organizational Performance is not a bolt on accessory, it is not a program, it is not a "go do," and it is not something you can simply purchase from some overpriced consultant and plug into your organization. Be very leery of those that would attempt to tell you (or sell you) otherwise. Human and Organizational Performance is a shift in our assumptions, it is a change in our beliefs, it's a new way of thinking about how we can improve the workplace, and it is ultimately a move from viewing people as a problem to manage, to viewing them as problem solvers. This journey is not easy, and it is not for the faint of heart, but it is well worth every ounce of work, struggle, and frustration along the way.

I call these concepts "ideas" in an effort to distance this conversation from these snake oil salesmen that insist that there is "one right way," that "right way" being their way of course, to bettering the safety of work. I call these concepts "ideas" in hopes that they are viewed as more of starting points, as vital areas of focus that deserve attention and deeper thought and exploration, and as the layers in between our underlying assumptions, principles, and the visible and felt work worlds in which we operate, rather than easy fixes or programs to manage. These "ideas" are just that, they are ideas. Ideas that, from my own experiences as a Human and Organizational

Performance practitioner, help us move closer towards doing safety better.

The ten ideas contained within this book were born out of years of learning (learning that continues each day) about the concepts of Human and Organizational Performance and Safety Differently. They are built with *the 5 Principles of Human and Organizational Performance* (Conklin, 2019) and the tenets of *Safety Differently* (Dekker, 2014) at their core:

The 5 Principles of Human and Organizational Performance (Conklin, 2019)

1. Error is Normal

2. Blame Fixes Nothing

3. Learning is Vital

4. Context Drives Behavior

5. How You Respond Matters

Key Concepts of Safety Differently (Dekker, 2014)

1. Workers are not the problem – They are the problem solvers.

2. We do not tell our organizations what to do – ask them what they need.

3. Safety is not the absence of accidents – it is the presence of capacity.

These ten ideas were developed, honed, and distilled through the coupling of these academic concepts and principles and their real-world application – these ideas emerged from practicing HOP and Safety Differently. With that being said, this book is geared towards just that, bringing Human and Organizational Performance to life within your particular organization.

I have tried to provide as many real world and "doable" items sprinkled throughout these ten ideas as I can, tactical bits that you can pick up and carry forward and build upon in your pursuit of safety better. I have purposely avoided being overly prescriptive, attempting to circumvent oversimplification and to dissuade these insidious "one size fits all" approaches to Human and Organizational Performance and the safety of work. As I have mentioned ad nauseum in my previous works, I am a firm believer in the old saying "there are a million right ways to do the same thing." I often state this concept by saying "there's more than one way to skin a cat." The intent of this book is not to prescribe to you "the ten right ways to skin a cat," nor is it intended to be a sole source guide to curing all the problems that you might encounter while skinning cats, nor should it be viewed as some linear tick and flick check sheet for skinning cats.

My hope is that these ideas, these lessons, and these learnings I have picked up along the way, that they will be of practical value to you as you embark (or continue) on your Human and Organizational Performance journey. I hope that these ideas encourage you to dig deeper than ever before, that they lead you to explore the numerous other bodies of work relating to Human and Organizational Performance and Safety Differently – the works that helped form my very own ideas as a HOP practitioner. I hope that this book encourages you to learn, to grow, and to challenge, that it gives you some ideas, tools, and a bit of confidence to go out and seek to render your work worlds better. My hope is that this book leaves you with more curiosity than ever before, that it leaves you excited to go out and try these ideas on for size, and that it leaves you excited about the potential future state of your organizations.

A shift in our views about people

As a HOP practitioner, and as a voice within the community of practice, I am frequently asked to define Human and Organizational Performance – to give my version of an "elevator pitch" – to provide a quick and clear description of the subject. To me, at the core of it all, Human and Organizational Performance is a fundamental shift in how we view people. It is the move away from viewing people as problems to be managed, and the shift towards viewing people as problem solvers. While there are several other vital bits and pieces, Human and Organizational Performance is about starting from a place of trust, embracing the human element of our work worlds, understanding that people show up to work to do a good job, and constantly and deliberately learning from those that do the actual work.

In our traditional approaches to the safety of work (and most other things for that matter) we often have started from a position of distrusting our fellow humans; we have viewed people as the source of problems and pain within our organizations. People have been viewed as the last great problem to fix, as the last step between us and safety utopia. We have viewed people as the problem to fix, and we seek to fix problems. We have built systems of distrust constructed of

24

endless rules, ones that are policed via mechanisms of constant surveillance, oversight, and harsh punishment for wrongdoers. We have tried and tried to comply and punish our way to safety excellence, but it has failed us time and time again.

Not only has our distrust of our fellow humans been a driving force for our mediocre (at best) approaches to the safety of work, but it has also been a harmful negative that has inflicted unnecessary pain and suffering upon those that diligently serve our organizations. This distrust of our fellow humans, and this desire to punish those "untrustworthy" and "uncaring" humans that we believe to be causal of our problems has led us away from safety, not closer to it. It has left our workforces fearful and untrusting, devoid of the ability to be honest with the organization and unable to tell "real deal" stories about how work normally occurs, and it has left our organizations blind to vital information and learnings.

The principles and concepts of Human and Organizational Performance moves us away from these misguided and harmful beliefs. Rather than viewing people as the problem and attempting to cure our work worlds of events and problems by seeking to cure people of their humanity, HOP teaches us to embrace our fellow humans, to defer to their expertise, to learn from them, to seek to understand, and to understand that their "know-how" and knowledge is vital to the success of our organizations. Human and Organizational Performance teaches us that error is normal, that no one chooses to make a mistake, that blame fixes nothing, and that blaming only moves us away from the so needed learnings we require to improve. Allow me to circle back to the key point, Human and Organizational Performance is a fundamental shift in how we view people – people are problem solvers, and we must create systems of trust so that they can do just that.

One of the first jabs at Human and Organizational Performance that many leaders take, especially those more comfortable with very vertical command and control styles of management, is that it is "too fluffy," "too squishy," or "too soft," but nothing could be further from the truth. As a close friend, one who is a high-level manager in the utility space, once shared with me, "Human and Organizational Performance lets me hear the raw and the real – that is what I need to make better decisions as a leader." Human and Organizational Performance is not about the squishy bits, it's not about being fluffy or soft, HOP is about getting down to the nitty gritty and digging into "raw and real" conversations and learnings. There is nothing squishy or soft about an employee sharing with you their near-death experiences.   There is nothing fluffy about hearing the story of a worker who amputated their finger but was forced to choose between reporting the event and getting medical attention or seeking medical care on their own so that they could keep their job.   The learnings and conversations that Human and Organizational Performance will bring about within your organization will be the rawest and realest conversations you have ever experienced.

Here is to embracing the raw and the real – I hope you enjoy the book!

DOING THE SAME OLD THINGS... OVER AND OVER

But safety is seemingly everywhere...

Are you stuck? Do you feel lost? Are your current approaches to the safety of work leaving you frustrated and without the "world-class" safety performance they once promised? What the heck is going on? How could this be? You are doing everything one could even possibly imagine as it relates to workplace safety!

You have written every procedure imaginable; you have drafted a rulebook thicker than the bible. You have even gone so far as to declare that injuries and accidents shall not occur. You have mandatory safety meetings each and every day, and every company gathering or meeting requires a mandatory "safety moment." You even send out daily safety messages and weekly safety newsletters to your employees, mandating that leaders take time to "stand down" and read them aloud to their crews. You are doing safety, a lot of it!

Upon entering your location, employees and visitors will immediately notice your efforts. Between the safety banners, posters, stickers, t-shirts, and hats, safety is seemingly everywhere. Employees and visitors alike are greeted with varying digital clocks, displays, and counters, each displaying "days since" an injury or some other specific occurrence of 'human error.' Even simple passersby cannot elude or ignore the presence of safety at your location, the gargantuan outdoor electronic display (strategically aimed at the nearby highway) proudly announcing "*2 million manhours worked without a recordable event*" makes sure of that. Even down to your average cup of morning coffee, you proudly drink from your company issued mug, its print announcing to the world "*ZERO is Possible!*" Safety appears to be ever present; a constant reminder to not be so foolish as to get hurt on the job.

But injuries are still happening. Not the scrapes, bumps, back sprains, or bee stings so much anymore, even close calls are a thing of the past, you rid your organization of those types of pesky occurrences years ago – you have gone 2 million manhours (and counting) without a recordable injury as mentioned already. The minor injuries almost never occur, but every so often, people are still getting seriously hurt or killed. Every now and again someone loses a finger, someone crushes their arm, someone gets mangled by a piece of equipment, someone gets horrifically maimed, or someone loses their life. The small events appear to no longer occur within your work world, but the big, bad, seriously not good things still happen every now and again. They happen unexpectedly, out of the blue, and without warning.

How could this be? You have made it to zero, you have rid your organization of all the lower-level events that your safety program (and industry) teaches as causal of fatalities and industrial maiming's, you have demanded compliance with your most sacred rules and procedures, and you quickly act to remove those "unsafe" and "uncaring" employees from your organization that do not strictly follow your most precious commandments. And, when these approaches fail you, you make damn sure that you are doing them harder, better, and with more focus and rigor than ever before. But nothing seems to work, nothing is getting better, and things often seem to be getting worse.

Safety is seemingly everywhere in appearance; all-encompassing and ever present. You, along with your leadership team, are constantly demanding compliance and surveilling the workforce to ensure they are meeting the requirements you have set forth. You are doing more safety, and doing it with more focus and rigor, with each passing day and event. But no matter how hard you try, no matter how much effort you pour into safety, nothing seems to be getting any better. What now?

A personal turning point…
Adapted from "WTFRM? A Reflection on What is Meaningful to Workplace Safety." S. Goodman 2021

I found myself in a similar situation some years back. I was only a few short years into a new role as a site safety manager for a large maintenance and construction contractor operating in commercial nuclear generation. I was frustrated, tired, growing more and more apathetic in my role as a safety practitioner, more and more nihilistic of our common approaches to worker safety, and increasingly depressed by our typical treatment of people within our work worlds. In fact, I was preparing to flee the safety profession, along with "safety focused" industries, altogether.

This specific memory or "turning point," one I often find myself horrifically flashing back to, starts on a warm and windy typical Arizona summer day. The sky above was clear and blue, it hadn't a speck of a single cloud. I can still hear the murmur of the surrounding crowd, I can still feel the sharp crunch of the gravel under my feet, I can still recall the smokey aroma of bar-b-que sizzling away on a nearby smoker, and I can still feel sweat dripping down my face from the hot sun pouring down upon my hardhat. This waking nightmare had not been born from some horrific workplace event or tragedy; everyone was very much alive and well. There had not been some explosion or meltdown, no one had been fired or ruined their career, it was actually quite the opposite. This pivotal and crippling moment was birthed from a celebration. The truly impactful bit of this experience grew more from the realization that the celebration brought to the surface, a realization that left me questioning nearly everything.

There had been no expense spared for this grand shindig. Trophies were purchased and engraved, thousands of t-shirts had been acquired, the caterers had been given a blank check, company executives from far and wide had descended upon the location, and we even brought in a professional DJ to "pump up the jams" as to "…give it that real party feel!" What had we accomplished that was so spectacular? What could possibly have led to such a grand gathering? Surely, we had done something BIG! We did do something… well, I guess the real answer to "why?" lies in what we didn't do.

We had just gone a year without a recordable injury. We had finally achieved the most sacred of safety accomplishments, we had finally cared enough, we had finally tried hard enough, and we were finally safe enough to get to ZERO. Our hard work had finally paid off and we were now reaping the rewards of entering into safety utopia. We had built an altar to zero, we built it out of heaps of "caring more" and buckets full of "trying harder." We built it on a foundation of sticks, carrots, and golden rules, and our unwavering commitment to our god Zero had finally led us to the safety promised land. We could now give out treasured safety relics, we could sacrifice hogs to consume their smoky fatty goodness, and we could do it all in the name of Zero. After years of the stick, we were finally rewarded with the golden carrot!

As I stood there for the cringe-worthy group photo, holding a shiny gold trophy in front of the *"Safety Excellence: ZERO is Possible!"* banner with the rest of the management team, I was overcome with dread and shame. Although in protest, there I was, smiling for the cameras and holding a relic to our safety god, Zero. "What a shit show…" I recall thinking to myself. At this time in my career, I had only recently been exposed to the ideas of Safety Differently. After years of frustration with more traditional approaches to safety, I was gifted a copy of Sidney Dekker's *Safety Differently* by a friend in the months prior to this event, it had set my mind on fire. I now found myself on a quest to learn everything that I could about doing Safety Better. But for years prior to discovering the works of Dekker, Conklin, and many, many more, I had been indoctrinated into traditional safety.

I was raised in a culture that viewed safety as an outcome to be managed, as a "bolt on" attachment (safety culture) that was separate from other bits of total organizational culture, an environment that viewed standdowns and increased observation or oversight as a way to manage people and their pesky behaviors, a work world that firmly believed that through this stick and carrot behavior modification process we could finally achieve zero, and that once zero was finally realized we would discover safety enlightenment.

As I stood there accepting an award for "dramatically lowering incident rates," as employees we're stuffed to the brim with pulled pork and ribs, as t-shirts and trophies were given away, as executives kissed babies and shook hands, as juicy bonuses were paid out, and as we all bowed and prayed to our god zero, an employee suffered a life-altering injury elsewhere at the location. How could this be? We had made it to zero, we had only had a small handful of minor scrapes and bumps, how, how, how? Of course, that handful of first aids were blamed, of course the involved employee was shamed and disciplined, of course they demanded greater leadership oversight along with bigger sticks and juicier carrots, of course the company chose to again double-down on the same old stale and near-useless traditional safety things, of course. But quitting was no longer an option, I had to make things better. I remember my thought clearly, a thought that has become a general rule for my career and my life, "I'll do everything in my power to make things better, or I'll get fired trying." This is a personal motto that I live by to this day.

The old "tried and true" had failed me yet again. But this time things were different; I was on a quest to do Safety Better. As I dug deeper and deeper, as I learned and explored more, as I experimented and innovated, as I worked to implement Human and Organizational Performance at various locations, as I wrote books and started podcasts, as I had conversation after conversation, I found myself coming back to a simple, yet extremely powerful question, "what actually matters?" What matters? What really, really matters a lot? Out of all the things that we seek to touch, impact, manipulate, and influence, what's actually important? What is trash and what is treasure? What is treasure that we have convinced ourselves is trash, and what is trash that we have painted gold and pretend to be treasure? What is real gold and what is fool's gold? How much of what we do is meaningful, how much of it is meaningless, and worse yet, how much of it causes more harm than good?

Clinging to our sacred cows

We've typically operated under a general rule of thumb that states, "we do safety better by doing more safety!" I'm nearly certain that you have felt this in your particular working environment. I'm confident that you have felt that constant buzz of questions around "what's next in safety?" There is a perpetual desire for new, new, new, but there is an equally matched desire to never part with the old or break away from the status quo. The idea of ceasing to do something, getting rid of a rule or safety slogan, or even replacing something with something better, is near blasphemy! We are safety junkies, addicts, and we want more and more, but only more of the same. We are constantly creating more useless bits and bobs for our so-called safety management systems to feed our addiction, a procedure here and a program there, we stuff in some new checklists and observation cards, and we are ever expanding and bloating the same old useless things. But have they worked? Herein lies a good bit of the problem; traditional safety has worked good enough.

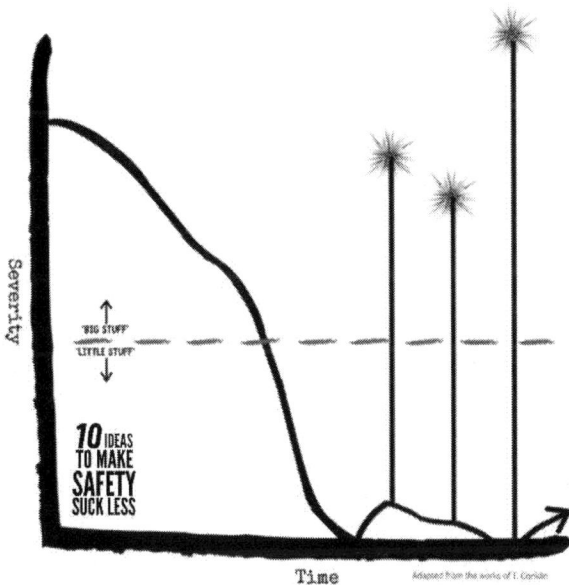

Over the years, our drive to prevent bad things from happening has indeed worked. More traditional approaches to worker safety have worked well enough for us to refuse to learn and grow beyond them. A quick examination of data around occupational injuries and fatalities will typically demonstrate a sharp decline in the total number of significant injuries and deaths over time. For many, if not most organizations, injuries have been reduced from regularly occurring excruciating maiming's and horrifying fatalities to a handful of bumps and scratches, with the horrific deaths and dismemberments now occurring with much less frequency (but remaining surprisingly consistent). At the very least, these types of events have become rarities for most companies, horrible anomalies that hide within our normal work. To add another layer of complexity to this problem, the sources of these rarely occurring and extremely high outcome events often only become obvious after their occurrence. If we could have imagined it, we would have done something about it. If we could have dreamed it up, we would have prevented it or reduced the severity of the event to more acceptable levels. We're often left holding a mixed bag full of somewhat frequent lower-level events such as bumps and bruises, and extremely infrequent high outcome events such as fatalities or life altering injuries. So, we seek to manage and manipulate what we think we can; we touch what we can see. We seek to excise the minor occurrences from our work worlds in hopes that they will somehow have a preventative impact on those more catastrophic events on the horizon.

There are a few key noteworthy "sacred cows" of traditional safety that are the likely culprits:

All incidents are preventable

And...

Closely examining and preventing small events allows us to predict and prevent big events on the horizon

So...

"Safe" means an absence of negative occurrences

And, because...

Most events are caused by "human error"

We should focus on human behavior because...

If people just followed the rules, nothing bad would happen

When shit eventually finds a fan, rather than honestly reflecting on the effectiveness of these deeply rooted beliefs, we double down on doing them harder. We write more rules, we preach to the frontline a harsher sermon about caring more, we measure and incentivize more, we beat the involved employees for not following the rules hard enough, we hold safety professionals and leaders accountable for failing to predict and prevent or for not oversighting hard enough, and we try to blame and shame our way to safety success. Is it any surprise that we can't seem to make any meaningful positive change in our workplaces? We continue to cling to these strange beliefs that we get better by doing the same things harder, if we finally rid our companies of bumps and scrapes then we will stop killing people, and if we finally fix people, all will be well.

The hamster wheel of safety insanity

As Albert Einstein so famously said, *"the definition of insanity is doing the same things over and over and expecting different results."* Our deeply rooted desires for comfort, predictability, stasis, and uniformity, along with our crippling fears of innovation, change, creativity, and unpredictability, has resulted in safety insanity within our work worlds. We cling to this insane state like a warm security blanket, even if it costs our employees their lives.

Why? Why do we persist in these beliefs even though we know that they are harmful? Simply put, it's two basic yet enormously powerful motivators:

On the surface, it all appears morally sound

AND

It's super easy

If we say something like, "no one should get hurt at work," that's a pretty morally sound statement, you'll get no argument from me. Now, when most of us hear a statement like that, we typically think something like this, "Yes, absolutely! No one should suffer a life altering event or die while simply trying to make a living." But that's not what this seemingly positive statement usually means to most organizations. It is often taken to the extreme meaning that says, "We shouldn't even have bumps, bruises, or scratches! We must achieve absolute zero or we suck at safety!" Why? All roads lead back to our underlying beliefs about what safety is, how we define "safe," and how we seek to influence it. If we genuinely believe that a.) *All incidents are preventable* and, b.) *we can prevent catastrophic events by managing and preventing lower-level events*, then that all makes great sense.

Unfortunately, life is not that easy. Although both of these key points make up the bedrock of most traditional safety management programs, they are near fairytale like illusions that only lead us astray and farther away from a focus on what actually matters. But, if taken at face value and not examined below the surface, it is easy to see how we lean into these ideas as morally superior and view them as a super easy way to approach safety.

The problem is twofold: 1) *Even though these traditional ideas continue to let us down, we insist that they will eventually work with enough focus and effort*, and 2) *They have worked well enough for us to be fearful or unwilling to part with them.* This is how we find ourselves stuck on the hamster wheel of safety insanity. After an event, whether it be a minor bump or bruise, or something far worse like an occupational fatality, we quickly uncover evidence to prop up these assumptions and beliefs. We scratch the surface with our investigatory efforts, easily unearthing numerous "should haves," "could haves," and "would haves," all pointing us in the direction of doubling-down on the use of our traditional approaches and tactics.

With ease we discover where increased oversight might have allowed for the prevention of the occurrence, areas in which the rules were bent or broken, that mistakes or errors were made by the involved employees, and on and on... With the gifts of hindsight and known outcome in one hand, and our traditional assumptions about safety in the other, we then swiftly point out things like, "If they just followed the rules harder," "If the leader had only been there," "if they only paid more attention or were more aware of the hazards," and the worst of them all, "If they only cared more."

Rather than challenging our assumptions about safety and the validity of our approaches to the safety of work, rather than digging deep into the complex sociotechnical systems in which work takes place, rather than undertaking a deep and meaningful examination of the context

44

surrounding normal work and the event, we often choose to continue on the path of safety insanity – we double down on doing the same old safety things, believing that if we just do them harder, better, faster, and with more rigor than ever before, then things will finally get better. We blame, shame, punish, and retrain our employees with even more frequency. We monitor, observe, and surveil our workforces harder than ever before. We bean count, measure, and trend to ever-growing extremes. We do it all – more and more of the same – harder and better than ever before, but nothing ever really gets better. We run and run as fast as we can possibly stand to run in our little hamster wheel of safety insanity – we run it to its breaking point. When it does finally break, we work quickly and diligently to build it back just as it was so we can hop back in for another round. Over and over, time after time, break down after break down, let down after let down, there we are, giving our traditional approaches to safety another spin.

We are living safety Groundhog Day

The early 1990's classic, starring the legendary Bill Murray, tells a tale eerily familiar of where we find ourselves with our approaches to the safety of work. Phil (Murray), a weatherman, is assigned yet again to report on the Groundhog Day festival in Punxsutawney. He is quite unhappy about the assignment. So, Phil spends his day in Punxsutawney doing what the film implies he does pretty much every year. Phil

45

spends his day mocking the festivities, belittling the people who partake in them, and generally assuming that the celebrations are beneath him. A blizzard blows in forcing Phil to have to spend another night in Punxsutawney. For reasons the film leaves unexplained, Phil wakes up the next morning to realize that it is February 2nd once again. He is doomed to repeat the day, caught in a time loop of unknown origin or duration, until, finally, he is able to live a day of selflessness, of joy, of love—and, therefore, to break through to February 3rd.

Aside from the filmmaking genius of the movie and Bill Murray's comedic brilliance, what can *Groundhog Day* teach us about our pursuit of safety better, about breaking free from our hamster wheel, about letting go of our repetitious use of safety insanity? While I could invest the entirety of this book into unpacking all of the deeply meaningful bits Groundhog Day touches upon, I will spare you that exploration. Rather, I want to hone in on this idea of breaking the endless circle, of stopping the endless loop by choosing to think about and approach things differently.

Time after time, Phil repeats his loop with no apparent end in sight. Initially he reacts with shock, confusion, and disbelief. Phil, now beginning to recognize that he is stuck in a time loop, devises a simple test to see if he is indeed living Groundhog Day on repeat. He retrieves a pencil, breaks it in half, and places it on his nightstand. However, when he wakes up the next day, the pencil is in one piece and has been placed back in the drawer where he originally found it. Still in shock and disbelief, now thinking that he is simply losing his mind, Phil seeks out psychiatric help – but nothing he does seems to help explain what is happening to him.

Phil's behavior quickly devolves into chaos. From police chases to suicide attempts, no matter what Phil seems to do, he is greeted every morning by his alarm clock radio playing the same song. He seems doomed to live out this time loop for all of eternity. How does Phil finally break the endless time loop he is stuck in? He finally chooses to do things differently – through Phil's purposeful choice to approach the situation differently he finally breaks the loop. Phil chose to do things differently and it gave him a different result.

We seem to be stuck in our very own version of *Groundhog Day*. In order to get a different result, for us to quit reliving our endless and nightmarish loop of safety insanity, we must finally choose to do things differently.   Doubling down on our same old approaches to safety, on our current assumptions about safety, on our beliefs about how we influence safety – even when we do those same old things with more effort and focus than ever before – only leaves us with the same results.

Evolving our Assumptions: Letting go and moving on

The only true path forward – the only place we can really start – is a targeting and reshaping of the underlying assumptions that have led to the present work worlds we find ourselves working to better. The great news is this: everything that has been constructed within your organization has been shaped or built by human hands. Everything that has been done can be undone, reshaped, reformed, bettered, or scrapped completely. The bad news? It is super hard and takes a really long time.

We have to shift our starting position; we must start from a place of better assumptions. These better assumptions, ones that we will touch on throughout this book, are vital in shifting our approaches to safety, and practically anything else we hope to do well. This shifting towards these better and more humanistic underlying beliefs that say people are the solution, learning is everything, safety is the presence of positives, error is normal, failure will happen, and on, and on, must become deeply engrained within us as individuals and within the organizations that we seek to positively influence. From this better starting position, we can begin to construct a better approach.

But I will warn you now, a healthy amount of this journey involves letting go and moving on from ideas, systems, processes, and industrial and organizational "sacred cows" that simply do not align with our better assumptions. It can be a painful and gut-wrenching undertaking for some organizations – holding on to all that they have built over the years with a death grip – fearful that any undoing or changing of their current systems will surely result in catastrophe.

This exercise can also get a little confusing from time to time. For me, personally, anytime that I find myself mumbling questions like "I don't know," "is this bad," "is this good," and on, I lean that much harder into the principles of *Human and Organizational Performance* and the tenets of *Safety Differently*. I always find the answers I am looking for – using the guiding principles as the lens. When you find yourself unsure of what to start doing, what to stop doing, or what to fix, firmly replant yourself into these principles we have already discussed – these principles along with the 10 ideas represented in this book (built with the principles of *Human and Organizational Performance* and the tenets of *Safety Differently* at their core) will not lead you astray.

Back to the point...

Doing safety better requires a reshaping and reformation of our shared underlying beliefs. If our desire is to create better and more humane work worlds, if we want to better the safety of work, if our goal is to make safety suck less, then that is where we must begin. We must set aside our desires for fixing surface-level symptoms, and we must dive much, much deeper into where our organizational problems grow from. We must replace flawed assumptions with better ones, we must have open and honest conversations about these beliefs and where they come from, and we must embrace better approaches.

While I wish I could say something like, "Just follow this super easy 12-step guide and world-class safety performance will be yours!" or "Just do X, Y, and Z, and you too can have Human and Organizational Performance at your company!" I can't. Wouldn't it be nice if we could just reduce it all down to some simple linear prescriptive method, one that guarantees consistent results, that we could all follow into the HOP promised land? But that will never be the case. Our work worlds are complex, messy, and chaotic living organisms.

With that being said, I feel that it is important to highlight how unique and complex your particular organization or workplace is. With this complexity in mind, I also feel the need to say that this is not meant to be some prescription you blindly take or administer to your organization. Rather than taking these ideas and attempting to force fit them into your work worlds, I hope that you walk away from this book with even more questions, and on a mission to learn. Not to learn from more and more safety or HOP theory (while that can be helpful), but with a thirst to go out and learn from those that GSD (Get Shit Done) within your organizations. Rather than prescribing to you "the one right way" to do Human and Organizational Performance, I hope that this book helps you along your journey – that it acts as a small piece of the puzzle. I hope that it provides you with a better lens in which to view safety and Human and Organizational Performance, and that it aids you in your pursuit of doing safety better. I hope that it helps you craft and mold better assumptions about how we approach people, learning, and the safety of work.

While I can't promise you some simple and easy fix to all that ails us in the world of safety and Human and Organizational Performance, I can promise you this: If we start with better assumptions, if we approach all that we seek to accomplish from a place of trusting people, we do things with people rather than to people, we ask better questions, and we create environments in which honesty is possible, things can only get better and work will suck less overall – and we will finally break free of doing the same old tired and ineffective things on repeat.

10 IDEAS TO MAKE SAFETY SUCK LESS

What now?

We have spent the first chapters of this book talking about our typical approaches to the safety of work, touched briefly on where these ideas come from, and discussed the ongoing safety insanity they bring about within our work worlds. How can we move our approaches to the safety of work in a better direction? How can we let go of these harmful and pain inducing "safety management techniques?" How can we evolve beyond these assumptions and beliefs that drive us to view people as problems to manage, to view blame as curative, these tactics that – no matter how hard we use them – continue to let us down?

A long hard look in the mirror

To begin, we must be willing to approach the problem head on. We must be willing to undergo the often-painful process of reflecting on our traditional approaches to the safety of work, to take a cold hard look at our assumptions about safety, to examine our assumptions about the people in our care, and we must be willing to drag the good, the bad, and the ugly out into the sunshine for thorough examination and scrutiny. We can no longer cling to our sacred safety cows; everything must be on the table. Be very cautious to not allow the most sacred tenets of our more traditional approaches to cloud your judgment or prevent you from progressing towards safety better. The most likely culprits that will prevent you from moving forward:

All incidents are preventable

Closely examining and preventing small events allows us to predict and prevent big events on the horizon

If people just followed the rules, nothing bad would happen

I bring up these few sacred tenants of traditional safety yet again due to the stranglehold they still have within our work worlds, and because of their ability to hold us back from growth. My intent is not to bash or beat up where we have been, it is to learn and grow beyond it. It is to acknowledge that these traditional ideas are not working, and in fact, they are making things much, much worse. I highlight these sacred cows yet again, due to their power to drive us to persist in our beliefs that if people only cared more, tried more, or did traditional safety harder, then all would finally be well. In order to do safety better, to better care for those we employ, and ultimately, to make safety suck less, we must be willing to finally part with these horrific ideas for good.

We move beyond bad ideas with better ideas. Acknowledging the good, the bad, and the ugly – staring it in the face and learning from it – is crucial to ensuring we do not continue on the hamster wheel of safety insanity.   But the continued introduction and pursuit of better ideas, that is how we truly move forward.

In the following chapters we will be exploring these 10 Ideas to make Safety Suck Less:

Idea #1 – Start from a Place of Trust

Idea #2 – Do Things with People

Idea #3 – Learn Deliberately and Often from those that GSD

Idea #4 – Pain Points are Starting Points

Idea #5 – Become Obsessed with the Things that (actually) Matter

Idea #6 – More Tools – Less Rules

Idea #7 – Stop Trying to Comply (or punish) Your Way to Excellence

Idea #8 – Redefine "Safe"

Idea #9 – Give up on Safety "Fortunetelling"

Idea #10 – Embrace Humanity

A quick note on order of importance…

While these *10 Ideas to make Safety Suck Less* are not laid out in some rigid prescriptive and hierarchical order (although they have been somewhat sorted), I would be remiss to not highlight a few good "starting points" for this journey – the "make or break" items that, at the very least, organizations must get right way more than they get them wrong. 1) *Start from a Place of Trust*, 2*) Do things with People*, and 3) *Learn Deliberately and Often from those that GSD* are a few of the most impactful starting points to pursue as you embark on this journey towards doing safety better.

While a solid case can be made that any one of the ideas on this list could be highlighted as "make or break," I highlight these three ideas in particular due to the profound impact they have on nearly every other effort or undertaking associated with doing safety better. These ideas focus on trusting those that we employ, letting go of our misguided parent-child approaches to management, and embracing deep and meaningful learning – learning from those that actually get things done. These ideas are the bedrock on which practically everything else is built. If we categorized all the other ideas in this list as ingredients for our safety better cake, these three items – starting from a place of trust, doing things with people, and learning deliberately and often from those that GSD – they are the mixing bowl. They are interconnected and interdependent. You will never find real learning without trust, you will never grow real trust without learning, you will never have either if you do things to people rather than with them – hand-in-hand seeking out safety better together.

While these three ideas are crucial, and I encourage you to start your journey by focusing your efforts on them, do not allow me to belittle the importance of all the others – each idea is vital to your pursuit. Many of these ideas have more traditional counterpoints or approaches associated with them. As an example, I highlight starting from a place of trust because our typical approach is to start from a place of distrust. I advise to do things with people because our traditional tactic is to do things to people, forcing our ideas of "what is best for them" down their throats. I hone in on embracing humanity because we have busied ourselves for decades trying to cure our workforces of it, rather than leaning into it. I could go on, but I will spare you – the traditional approaches each of these ideas are designed to move us past are quite obvious. Allow me to circle back to the point, each of these areas carry significance in evolving us beyond our starchy, stale, harmful, and inhumane approaches to the safety of our work worlds.

With that being said, it is vital for you to understand where your organization is presently at. Seeking to understand where you are at in the present moment will allow you to craft a bespoke approach for your journey, rather than seeking out some misguided and ineffective copy and paste or "off the shelf" approach to Human and Organizational Performance.

Where are you?

To bring us back to *the 5 Principles of Human and Organizational Performance* (Conklin, 2019) – Learning is Vital. I have adapted this for the purposes of this book by creating the actionable statement of "Learn Deliberately and Often from those that GSD (get shit done)." That is exactly how you will discover where your organization is currently at – you must go out and learn where you are at. You must seek to understand where you are currently at by listening and learning from those that live within your particular work world.

When helping organizations initially start down this path, and at various "check and adjust" points while on their journeys, I have found the use of learning explorations or listening sessions very beneficial to understanding the reality of the situation – and that is exactly what we should be in search of, reality. Not where you think you are at, not where the executives believe the organization to be, not where managers insist you are, you are seeking out raw and real information on the reality of the situation. You are seeking out vital operational intelligence from those that spend their days working and getting things done within your organization.

I refer to these approaches as learning explorations, or as "freestyle" methods to operational intelligence, as there is often not a specific problem at hand to solve or explore. As compared to the more normal use of learning teams or other operational learning tactics that normally start with a particular problem or pain point to work on bettering, these sessions typically focus on very broad inquiry using conversation starters such as:

What are we really good at?

What are we great at?

Where could we do better?

Where are things more challenging than they should be?

What should we start doing?

What should we stop doing?

And other similarly broad questions designed to explore the present reality of our work worlds.

These approaches are used to unearth starting points for further learning, they are used to help you discover the problems at hand and current challenges faced by the organization, and they will help you craft and prioritize a custom-tailored approach based on what you learn.

If your organization is already venturing down the path of Human and Organizational Performance, you can easily include more focused conversation starters relating to your organizations progress such as:

We have been focusing on doing things differently, what kind of changes have you noticed?

Has anything changed (for better or worse) in your day-to-day job?

How do you feel about this new approach the organization is taking?

Again, these broad questions should be designed to dive deep into where the organization is currently at and explore the lived experiences of employees working within the organization.

Pretty broad right? You would be correct in thinking that we are casting quite a wide net here. But that is kind of the point with these macro approaches to operational intelligence – we are not seeking to solve a particular issue or problem – we are seeking raw and real information relating to where we are at and what we should focus on. To do that requires the casting of a large net, it requires broad non-specific approaches to seeking out "real deal" information, it requires a healthy dose of chaos in our approaches to searching for the raw and real stories of our employees. Too much structure with these particular approaches to operational intelligence, too much rigidity, too much order, or too narrow of a net, would result in missing vital stories, experiences, and crucial pieces of valuable information. Embrace the chaos, embrace raw and real conversations, and lean into them, if you truly seek to learn and understand.

A few tactical bits and tips relating to these sessions

You will be left with page after page of notes and information from these sessions. Personally, I prefer to leave my clipboard in the car, opting for the use of flipcharts during these conversations. As we work through the sessions, I find myself feverishly capturing information on these flipcharts and plastering them all over the walls of wherever we are meeting. This "wall of discovery" allows for participants (and the facilitator) to easily see all that has been shared, to highlight and prioritize particular items, and to group certain related things together.

Do not kill these meetings by appearing to be buried in your notebook or clipboard – do not be so focused on capturing every minute detail that you stifle the conversation altogether. You will have plenty of time after these sessions to digest everything you have heard and to pull together your own personal notes. As mentioned, I find it easiest to capture everything on flipcharts, opting to take a picture of our "wall of discovery" after completing the session. I find myself going back to these pictures time and time again throughout the process of distilling down information and while crafting a path forward. I find myself pulling out various high priority items, opportunities for "quick wins," and circling and highlighting various items that we need to learn more about.

I do not want to get overly prescriptive in how you approach these conversations, in fact I often depart from more common approaches to learning teams with these particular sessions (more to come on learning explorations later in this book). Typical learning teams are conducted in five basic steps:

1. **Prepare** – Select around 6 people close to the work, mix them together and have a conversation.

2. **Learn** – In this first session discuss and discover how work actually gets done.

3. **Soak** – Do you have those 2 AM "ah-ha" moments like I do? That's the point of this breaktime. This time allows for information to soak in and for ideas to bubble up to the surface.

4. **Improve** – In this second session you will review what was discussed in the last session, but now the conversation moves towards ideas on how we can improve.

5. **Action** – Now we turn these ideas into actions that solve problems, add defenses, remove error traps, and make things better.

With these more "freestyle" learning explorations, I approach them with even less structure than indicated above – often they are single sessions, and we are not taking "soak time" or diving into focused actions. My point here is this, do what works. Be prepared, keep the groups to a manageable size, create ways to follow up or for participants to get their 2:00 AM "ah-ha" moments to you, and build a little structure to your approach, but keep it loose and natural and figure out what gives you the best results. Avoid getting mired down in too many steps, too much structure, or too many rules.

Another common question I get when being asked to guide organizations through this initial pulsing of reality is "how many people should we involve in these sessions?" I often reply with a healthy dose of my naturally occurring sarcasm by saying something like "enough." But there is quite a bit of truth in my answer. It's far too simplistic to pick a random number or percentage out of thin air – these magic percentages are quackery at best anyways. How about this, "a good chunk of them" or "hell I don't know." In all honesty, I typically do not target certain percentages or certain numbers of employees. I keep going and going until I have a good understanding of where the organization is currently at. You should be seeking balance here; you have limited resources and people. When your learnings begin to get overly repetitive, it is a good sign that you are reaching the end. Also, it is based off of what is realistic. If your organization employs 10 people, you should probably talk to 10 people. If your organization employs 10,000 people, talking to everyone is going to be quite an unrealistic undertaking. We find ourselves back at my original answer of "enough." We also find ourselves back to some of my earlier points on not being too rigid in your approach or overcomplicating things. Do not waste your precious time obsessing about the minutia – jump in and start learning – you will typically discover "enough" somewhere along the way.

A quick note on the structure of the following chapters

I have tried my best to craft the following chapters in a manner in which they can be read as standalone pieces. Feel free to skip around, go where your curiosity leads you, or flip to areas in which you are seeking out particular information for your efforts. While I feel that each chapter contains relevant and useful information, no matter where you find yourself on your Human and Organizational Performance journey, feel free to follow your curiosity.

I also encourage you to dig through the resources section at the end of the book. I have listed many great books, podcasts, websites, and people that I am certain will be immensely valuable to you in your Human and Organizational Performance endeavors.

And finally… 10 Ideas to Make Safety Suck Less

With all of the proper introductory bits out of the way, with a good understanding of how to learn about where your organization is currently at, and with some thoughts provided on how to prioritize your approach, lets dive right into the meat of it.

START FROM A PLACE OF TRUST

CONVERSATION PRIMERS

Are your organizational systems built on an assumption of trusting employees or distrusting employees?

How regularly does your company seek out blame when things do not go according to plan?

What types of questions are asked after an injury or event?

10 IDEAS TO MAKE SAFETY SUCK LESS

…rather than from a place of distrust.

The company requires you to upload an itemized receipt for that cup of coffee you purchased on your last work trip into their expense resolution system, and they require that you keep a hard copy until the end of time; your boss demands a thorough accounting of how you spent each minute of the previous week be emailed to them every Friday afternoon for their review and critique; the company bans the carrying of pocketknives at their locations; your organization applies GPS monitoring systems to practically anything with wheels; a strict "professional" dress code is enforced by the company that covers everything from types and styles of haircuts, to the general statement of "all employees shall wear pants at all times while at company locations."

We reside within work worlds built upon the bedrock of distrust. Surely, we cannot trust a person to buy a cup of coffee without undergoing a thorough auditing of the purchase! We could never possibly trust a seasoned and high-performing professional with something as critical as managing their own time and priorities! Although we trust people with multi-million-dollar pieces of equipment, the management of wildly complex processes, and to work with insanely hazardous things, how could we ever trust them enough to allow them to carry pocketknives? How could we ever trust them to drive a car? How could we ever trust them to wear trousers?

Simply put, we infantilize our highly skilled and knowledgeable workforces. We lean heavily into this misguided notion of "management knows best, always!" We honestly think that we know what is best for them, that we must do "what is best for them" to them, that we must do it whether they like it or not, and that we require little to no input from them because surely, they could not possibly know what is best for them. As sad as it is, we simply cannot muster up the ability to trust the people we employ. We write rules, micromanage, surveil, monitor, and brutally punish those unlucky few that we find to be out of compliance with our Tayloristic and parent-child approaches to overseeing our workforces.

Often, the only "wrongdoing" discovered is non-compliance with the surveillance mechanism or management system itself – a missed form, forgetting to upload a receipt, the avoidance or bypassing of vehicle monitoring systems, and various other acts that many organizations believe to be near-treasonous offenses. The targeted behavior is not even caught, simple non-compliance with the system or process put in place to catch the behavior is enough to warrant extreme corrective measures taken against the offender.

We genuinely believe that if we do not have these rigid structures of rules and monitoring then our work worlds will devolve into chaos – we believe that these mechanisms are strong defenses that prevent undesirable behaviors from manifesting in our workplaces. We just cannot seem to see beyond these simplistic, misguided, and ineffective approaches. Tactics that regularly harm the majority of our workforces – the people that would never purposely seek to take advantage of or cause harm to the organization – while almost never catching the miniscule amount of people that actively seek to take advantage of or cause harm to the organization.

We are writing rules, building large *Orwellian* monitoring systems, and using brutal enforcement tactics in the hopes of catching "wrongdoers," but all these systems are catching (and harshly punishing) are honest hardworking people trying to get work done within a complex and everchanging world. We invest massive amounts of time, energy, and resources into constructing these systems, sometimes even creating entire departments dedicated to them, all in an attempt to catch those that would dare to toe the waters of non-compliance.

Sometimes we even go so far as to purposefully set up meticulously camouflaged and well laid traps within our work worlds. Pretending like we are trying to snare rabbits or hunt big game; we lay in wait – treating our employees as if they are some form of game animal – to test their willingness and ability to comply. We monitor, evaluate, and question every action (or inaction) they take while attempting to get things done. We do all of this and so much more, all in the name of distrust.

Distrust and blame go hand-in-hand

Distrust and blame are like peanut butter and jelly, bacon and eggs, toast and beans (for my English friends), or biscuits and gravy (for my southern friends). They are the perfect marriage of our human desires, coupling together our innate instinct of distrust and the "feel-good" exercise of blaming into a vengeful monster we then set loose upon those we feel have wronged our organizations. Blame is easy, it feels good, and it makes us really feel like we are taking the moral high ground by harshly punishing others that would break the rules – something, we of course, would never dare to do.

We really favor blame in the space of employee safety. Blame is well primed due to the untrusting views already held about workers. These views are that much more untrusting when related to something as serious as safety and health. Adding fuel to the fire, most organizations view safety as a "you" based activity, as something you choose to get right (or wrong), as a simple endeavor that only requires enough attention, care, and focus, applied by the end user to get right, and to avoid the occurrence of events. So, organizations see the application of blame as the obvious choice for most safety and health related missteps or events. Many organizations never trusted their workers to get safety right in the first place – hence their massive structures of rules, surveillance, and enforcement – so when a safety event eventually happens, the end user of the organizations safety systems is swiftly blamed, shamed, retrained, or worse.

We will then drag out our long list of "you" based safety platitudes. We will screech things like "you should have paid more attention" "you should have done a better hazard analysis," "you should have been more responsible for your own safety!" We will look back on these horrific statements as causal of the event at hand and swiftly land on blaming the involved worker.

Possibly worse yet, we use the misguided beliefs as evidence for the need of even larger and stricter systems of distrust – even larger and harsher structures of rules, surveillance, and enforcement. On and on we go, deepening the divide between the organization and the people it employs, and ever strengthening the parent-child relationship we have built with the workforce.

The infantilizing of our workforces

Our primary position of distrust has driven us to treat employees like they are unruly and rambunctious school children, or like they are rebellious and defiant teenagers. This desire to treat our employees as if they are children has only seemed to grow in recent years – we have evolved to become proverbial "helicopter parents" to our workforces. I am nearly certain that you know the constantly hovering, persistently monitoring, and the 'ever ready to leap into action' type of parenting style I am referencing – a style of parenting characterized by over focus. Ann Dunnewold, Ph.D., a licensed psychologist and author, described the phenomenon of helicopter parenting in a 2019 *Parents* article as "being involved in a child's life in a way that is overcontrolling, overprotecting, and over perfecting, in a way that is in excess of responsible parenting" (Bayless, 2019). In research from McCarthy and More (2021), helicopter parents tend to:

- *Worry about safety*
- *Place heavy restrictions on what children can and cannot do*
- *Swoop in to solve problems for children who can likely solve the problem themselves*
- *Impose constant supervision and correction*
- *Make decisions for their children without any input from them*
- *Overly involve themselves with children's teachers and coaches*
- *Keep lines of communication with the child constant, zero independence from one another*
- *Have some level of anxiety or fear*
- *Refuse to allow failure as part of the learning process*

Helicopter parenting can have rather dire consequences such as decreased self-confidence, diminished self-esteem, development of entitlement, anxiety and depression, and the development of hostility towards parents for maintaining extreme control over their lives and their decisions (McCarthy & More, 2021).

Parents tend to gravitate towards this overbearing style of parenting due to fear of consequences, anxiety, overcompensation, and pressure from the outside world – these are often the primary reasons why parents go into full-blown helicopter mode (McCarthy & More, 2021).

Is it all beginning to sound a bit too familiar? Not only have we infantilized our workforces through the application of parent-child approaches to management, but our organizations have gone into "full-blown helicopter management mode." We hover, we monitor, we constantly coach, correct, and micromanage, and we are creating the same negative consequences brought about by this overbearing style of parenting. The problem deepens – our employees are not children. Our employees are not our children, but we treat them as if they are. We fall into this trap for many of the same reasons that parents do – we fear the consequences of not hovering, we constantly monitor to curb our anxiety, we seek absolute control in hopes of steering clear of potentially dire consequences, and because we see so many of our peers and competitors doing the same. But the fact remains, our employees are not children, and continuing to treat them as if they are, only serves to create harm and vast unintended negative consequences.

Some obvious consequences of infantilizing our workforces:

- *Reinforcement and solidification of a parent-child relationship with employees*
- *The creation of an "us v. them" atmosphere*
- *Less openness and honesty*
- *Less "raw and real" conversations*
- *Victimization of the workforce*
- *Vilification of the organization*
- *Degradation of ownership and accountability*
- *Vast amounts of time spent hiding or covering up behaviors*
- *Less engagement*
- *An undermining of skill and wellbeing*

The negative side effects of our infantilizing approaches to the management of our workforces seems nearly endless. These negative and often unintended consequences are brought about by coupling our desire for blame with our parenting-like approaches to management, and then using our normal torture kit of blunt instruments (like disciplinary action), all in some misguided attempt at creating positive influence and outcomes within our work worlds. But they never work out, they never work as we intend for them to, and they only serve to harm the workforce while leaving the organization blind to vital operational information and with a false sense of security – one that says, "all looks well from here."

Shifting our assumptions about people

The Oxford Dictionary defines trust as "*a firm belief in the reliability, truth, ability, or strength of someone or something.*" Personally, I find it very interesting (and feel the need to highlight) that nearly every mechanism we use during pre-hiring and onboarding processes are exercises designed to develop our trust of those we are seeking to employ. We ask situational questions during interviews, we perform background screenings, drug testing, nicotine testing, skill assessments, practical skill evaluations, and reference checks, all in the name of seeking out trust. In more extreme examples we use things like lie-detector testing and psychological tests, examinations such as the polygraph or the MMPI (Minnesota Multiphasic Personality Inventory – a psychological test that assesses personality traits and psychopathology), all in hopes of developing trust in the applicant – these are all explorations into the trustworthiness of those we are bringing into our organizations. We invest vast amounts of time, energy, and money into ensuring that we are hiring high-performing and reputable individuals. Yet, as soon as we bring them into our work worlds, we submerge them into our systems of distrust. Even after this barrage of pre-employment poking and prodding, we still distrust those that we choose to employ.

Organizationally speaking, we have a low propensity to trust. We start with poor assumptions about those that make up our workforces, and we then point to some small handful of more extreme negative events to prop up our logic for maintaining this untrusting position. We'll quickly point out an example where an employee was caught embezzling, that one was found to be charging personal expenses to their company credit card, we will highlight that an employee once cut themselves with a pocketknife and required a trip to the hospital, that a person was once discovered to be drinking on the job, that someone once stole some private customer information – we will use these and other similar examples as our reasoning for continuing to distrust our employees. We use these rare events to form the basic assumption that people should not be trusted, and we broadly apply this belief to all those that we employ. Some basic assumptions we draw about those that we employ:

- *We simply cannot trust people to do the right things*
- *People are always trying to "get something over on the company"*
- *They lack integrity*
- *They avoid responsibility*

- *People rarely act with good intentions or with the company's best interests in mind*
- *Without constant supervision and monitoring, people will be less productive, less safe, take greater risks, break the rules, etc.*
- *They are fundamentally lazy and desire to work as little as possible*
- *And more...*

These assumptions form the foundations of our systems of distrust – they lead to the artifacts of distrust that we can visibly see or experience within our work worlds. To move beyond our systems of distrust, to embrace trust as our organizational neutral position, these basic assumptions must be reformed and reshaped into better assumptions. Without a fundamental shift in how we view those that work within our organizations, almost nothing will change.

Our new normal – trust as the organizations neutral position

To shift these assumptions, we must lean into the *5 Principles of Human and Organizational Performance* (Conklin, 2019), and the tenets of *Safety Differently* (Dekker, 2014). We must genuinely shift our assumptions towards viewing people as the solution rather than the problem, we must grow an understanding that error is normal and attempting to punish people into not making mistakes only creates harm and undermines learning, we must lean into better assumptions that tell us:

- *Most people only want to do a good job*
- *People want the organization to succeed*
- *They have integrity*
- *They are responsible*
- *They are highly skilled at what they do*
- *They care, a lot*
- *And more...*

This shift in assumptions will move us beyond our desires for blame, push us to seek out restoration over retribution, and drive us to deconstruct our systems of distrust. The time, energy, and resources that are currently consumed by our mechanisms of monitoring and surveillance can be better spent on asking employees what they need to be successful, and then providing them with just that – the things that help them rather than hurt them.

Trust... even when shit hits the fan

Trust can sometimes feel a bit easier to focus on when things are going well, but its continuation when things have gone awry is vitally important to overcoming our leanings toward retribution, poor reactions, and other problematic items that discourage or prevent us from learning raw and real information about operational surprises occurring in our work worlds – information that is crucial to making better operational decisions and growing betterment within our organizations.

We must purposefully exercise trust when we encounter these surprises by leaning heavily into the better assumptions we have already discussed. When that not-so-great something does happen, when there is an operational upset, a quality escape, an injury, or worse, we must start from a position of trust. A few better assumptions to apply in these situations:

- *Employees do not choose to make mistakes*
- *Everything made perfect sense to those doing the work, until it suddenly didn't*
- *If they knew this was going to be the outcome, they would have not proceeded*
- *They made the best possible decisions they could with the information they had at hand*

Focus on seeking restoration

We can begin to move towards restoration by letting go of our typical investigative processes, ones that commonly mirror criminal investigations and focus in on things such as rule violations, the gathering of witness statements, and the collection of evidence. To begin, we can start by asking better and more meaningful questions. According to Dekker (2016), a restorative approach to organizational justice asks questions such as:

Who is hurt?

What do they need?

Whose obligation is it to meet those needs?

This focus on restoration is in stark contrast to our typical focus on retribution, one that often leaves us asking questions such as (adapted from Dekker, 2016):

What rule was broken?

How badly was it broken (or bent)?

Based on the above, what does the "wrongdoer" deserve?

In more recent iterations of retributive approaches to organizational justice, I have seen these lines of inquiry take a softer turn, but the focus remains the same – who broke the rule and what do they deserve? We will ask additional things like:

Was the rule one of our "rules to live by?"

Did they know about the rule?

Were they trained on the rule?

Was this a willful violation, unintentional error, common mistake, and etc.?

But are we really asking anything all that different? Not really. We are still seeking out opportunities for blame, seeking out organizational sins, and chomping at the bit to swiftly "hold people accountable" through the extraction of flesh from those discovered to be so foolish as to violate our most sacred rules. But where has that gotten us so far? Sure, discovering a so-called "violator" or "rule breaker" feels good – it feels like the right thing to do, and it eases our anxiety by making us feel like we have solved the problem. But nothing has been learned, nothing has been made better, and work has not been rendered "safer" through our pursuit of blame and punishment. In fact, a strong case can be made that our efforts are doing just the opposite of what we intended them to do.

The application of blame and punishment within our work worlds does quite a bit, it just does not do what we think it does. We think that we are making our workplaces a bit safer by the removal of pesky and uncaring individuals, we feel that we are teaching people vital lessons through the purposeful application of pain and suffering, we believe that we are demonstrating to our employees the consequences of bending or breaking the rules by making harsh examples out of those that do, and we have genuinely convinced ourselves that we will (eventually) punish our way to excellence. So, what actually happens because of our focus on retribution? Absolute silence – silence that is only broken when a failure is so large that it cannot be hidden away.

For the purposes of this book, a focusing in on the more tactical applications of Human and Organizational Performance within organizations, I want to direct attention back to the HOP principle of "learning is vital" (Conklin, 2019). To learn or to blame, is a choice we must actively make as organizations – a choice between two mutually exclusive paths forward. Taking the path of blame is to actively choose to forgo learning. This is where we find ourselves back to embracing better assumptions – choosing to start from a better position even when things have gone wrong – and understanding that choosing to learn less (if at all) by seeking out blame does not serve to make us organizationally smarter. When the not-so-great things occur (and they will), we must respond with a focus on restoration by asking better questions – who is hurt? What do they need? Who is responsible for getting them what they need? Is the location safe and secure? If it is not safe and secure, how can we render it safe and secure? As we move beyond our initial response to an event, we must focus in on raw and real learning – we must seek to understand.

Seeking to understand

To continue down this path of "tactical applications" of Human and Organizational Performance, let's dig into some better approaches to learning from events. When an unintended operational upset occurs, whether it is an injury, an environmental event, or quality issue, we must seek to understand. Based off the better post-event assumptions we have already discussed; we must purposefully and deliberately take a more learning centric approach. In contrast to our more traditional approaches to investigating events, ones that have focused heavily on individual behaviors and errors, our more learning centric approaches take a deep dive into normal work, into the context surrounding the occurrence, and seek to learn enough that we can understand or "put ourselves in the shoes of" those that experienced the event. These efforts are collaborative endeavors – avoiding offenders and prosecutors – that involve and learn from those nearest to the event, rather than placing them on trial.

Some final words on starting from a place of trusting our fellow humans

Of course, we can choose to continue to be distrusting of our fellow humans. We can choose to continue to maintain and operate our systems of distrust – ever trying to punish and comply our ways to operational excellence. But allow me to insert a rather pointed question here: What kind of existence is that? What type of awful dystopian future awaits us within our work worlds (and beyond) if we continue to embrace these misanthropic views of our fellow humans – especially for those that are entrusted to our care?

If our hope is to craft a better work world (and, a better world in general), then we must invest our efforts into building and maintaining systems of trust rather than distrust. If our hope is to do things well, we must embrace and lean into trusting our fellow human beings.

KEY POINTS

Our current work worlds are built upon the bedrock of distrust

We really favor blame, especially in the space of employee safety

The application of blame and punishment within our work worlds does quite a bit, it just does not do what we think it does

Trust must become the organizations 'neutral' position

We must embrace and lean into trusting our fellow human beings if we desire learning and betterment

PUTTING IT INTO
PRACTICE

- Focus on trusting down through the organization rather than asking for employees to trust up – genuinely extend trust to those that do the work
- Shift organizational approaches and processes towards a focus on restorative justice
- Seek to eliminate the petty and harmful systems of distrust throughout your organization

10 IDEAS TO MAKE SAFETY SUCK LESS

DO THINGS WITH PEOPLE

CONVERSATION PRIMERS

How do you currently seek to bring about change within your organization?

What part do your employees play in helping to steer the direction of the organization?

Do you involve employees in the creation of the organizations processes, procedures, and programs?

10 IDEAS TO MAKE SAFETY SUCK LESS

…rather than to people.

It seems to be deeply embedded within our nature – this telling people what to do – so much that we just cannot seem to help ourselves. From what to eat, what to wear, how to be happy, who you should vote for, and yes, even to the safety of work – we all have an opinion on what is good for others, and we must share it. In our minds, and based off our lived experiences, we truly believe that we know what is best, not only for ourselves, but also for everyone else. We really like to do things to people – giving them often stern and rigid advice – and we expect them to do what we say.

I would be remiss in not stating that some of this is likely brought on by a yearning to exert some sense of moral superiority. I would be a bit derelict in my duties, if I did not state that some of this potentially comes from our longing to be right, and to prove others wrong. I would be foolish if I did not note that these items likely play a role in forming this inclination to tell others what to do, and that they likely lead us to try to do things to people. But, in my opinion, I feel that this desire is often birthed out of care, out of hopes of being helpful to those around us, or from a desire to protect people.

The problems?

Problem #1 – *People really do not like being told what to do.* Remember that time when you were at work on one of those hot sticky summer days – that time when your boss "corrected" you for pulling off your safety glasses for a few seconds so they could defog? What about that time when you glanced at your phone while driving only to receive a snappy "don't do that!" from the passenger's seat? Or maybe that time when you told your significant other that you were committing to losing a few pounds, but when you suggested taking a day off from the gym you were met with "I thought you were losing some weight? Go to the gym!" How did those situations make you feel? Often, those types of encounters (being told what to do or what is best for you) leads to feelings of anger. Scientists have termed this phenomenon psychological reactance. Psychological reactance is our brain's response to a threat to our freedom. Threats to freedom include any time someone suggests or makes you do something (Hall, 2019).

We react to these threats in several different ways – through actual actions (or inactions) that extend well beyond our internal thoughts and feelings. We rebel, in hopes of restoring or reasserting our freedom; we comply, deciding to like and accept the prescribed action; or we ignore, choosing to pretend that the advice or direction was never given in the first place.

Rebel

Our telling people what to do – our doing things to people – really backfires on us with this particular response. Often, we purposefully do the exact opposite of the advice or the direction that is being given to us. We seek to maintain or restore our freedom through our rebellion, even when the advice or direction is generally good for us – even when we know what we are doing is potentially self-harming. Health communication experts note that this rebellion sometimes occurs in response to campaigns that tell people to quit smoking. Rather than reducing smoking, these ads sometimes cause people to want to smoke more (Hall, 2019). When we feel that our choices are being restricted, or that others are telling us what to do, we sometimes rebel and do the opposite.

Comply

An individual choosing to comply is the desired outcome we typically seek when prescribing to others what they should or should not be doing. In fact, what we are often seeking is obedience – compliance and obedience being two entirely different things. Unlike obedience, in which the individual making the request for change is in a position of authority, compliance does not rely on a power differential (Cherry, 2022). Compliance involves changing your behavior because someone asked you to do so. While you may have had the option to refuse the request, you chose to comply (Cherry, 2022). From our own personal experiences, we can easily recognize that people generally only tend to comply when they believe the advice to be well founded – when they feel the direction being given is good, agreeable, and sensical. According to research from Cullum et. al (2012), the presence of these factors makes it more likely that people will comply:

- *Affinity*: People are more likely to comply when they believe they share something in common with the person making the request.
- *Group influence*: Being in the immediate presence of a group makes compliance more likely.
- *Group size*: The likelihood of compliance increases with the number of people present.
- *Group affiliation*: When group affiliation is important to people, they are more likely to comply with social pressure.

When we decide that we like the advice or direction being given – often because we either do not see it as a threat to our freedom, because the advice is beneficial enough that it outweighs the threat to our freedom, or because the threat to our freedom has been minimized – we might just choose to comply with it.

Ignore

In this particular response to psychological reactance, we deny that the threat to our freedom ever existed. Think of it like this, imagine that you are a few plates into dinner at your favorite all-you-can-eat buffet when one of your in-laws' comments to you, "going for another helping I see..." I can only guess whether you would choose to rebel, comply, or ignore, but I will be going into full-blown ignoring mode. I pretend that the comment was not made, I ignore the advice, and I go for a fourth helping of sesame chicken, even though I know it is not good for me.

Problem #2 – *We do not understand* – we really think that we do – but we simply cannot. The behaviors of others seem so very simple to those not "walking in their shoes." In a 2017 *Chicago Booth Review* article, Alice Walton sums this up nicely stating that *"people tend to think they can understand others simply by watching them—but they can't read people as well as they think. Understanding another person actually requires getting perspective by being in his or her situation."* We simply do not possess the ability to truly walk in the shoes of another, we do not have access to their personal experiences, feelings, emotions, perspectives, or thoughts.

Highlighting this inability to understand – focusing primarily on our inability to understand another's emotional state – scientific research suggests that people have too much confidence in their ability to read people, and that they are often painfully unaware of this overconfidence (Zhou, Majka, Epley, 2017). Heidi Grant, in an article for *Harvard Business Review* (2015), highlights studies of college roommates – examining to see if over time your roommate was more likely to begin to see you the way you see yourself – in which almost 400 college roommates were tasked with describing their own personality along with their roommates. The findings were quite interesting, revealing that it took nearly nine months to even begin to get in sync (Grant, 2015). Grant additionally notes that, even after nine months, the correlations between how college students saw themselves and how their roommates saw them were surprisingly low.

Let's take this a step further by inserting the element of cognitive bias. Specifically highlighting fundamental attribution error, or our tendency to attribute another's actions to their character or personality, while attributing our own behavior to external situational factors outside of our control (Mcleod, 2018). In other words, we tend to assume that a person's actions depend on what "kind" of person that person is, rather than on the social and environmental forces that influence the person. We tend to see others as internally motivated and responsible for their behavior. When we see things that we seriously do not like – especially things relating to safety like behaviors that are viewed as dangerous or risky – we are quick to snap into judgment mode. We tend to label those we see exhibiting these "risky behaviors" or breaking the rules as uncaring, irresponsible, dense, or lacking proper hazard awareness, and we swiftly justify these labels by highlighting how we would never consider doing something so foolish. We land on what we believe to be the problem at hand – a bad person.

Where is he going with this...

We seem pretty terrible at understanding each other along with the vast forces that are at play in shaping behavior. To make matters worse, we tend to think that we are pretty great at understanding those around us and how to shape their behaviors. On top of that, we tend to like to assign intent to the actions of others – leading us to view them as bad people. Due to these inaccurate and misguided beliefs – ones that tell us things like we do understand, we do know what is best, we need to fix bad people – we attempt to do things to our workforces, especially with safety related things. Unfortunately, this (often extreme) effort to do safety to people has unintended consequences such as driving people to rebel or ignore.

If we cannot understand those that we seek to influence, how can we seek to influence them? The Principles of Human and Organizational Performance (Conklin, 2019) tells us that context drives behavior – a key concept of Safety Differently (Dekker, 2014) says that we should not tell our organizations what to do, that we should ask them what they need. We must fully embrace these ideas. Our desires to manage and manipulate employee behaviors, simply by asking people to change their behaviors, never really works out well for us. As my friend Clive Lloyd states in his recent book, *"behaviors are not the problem – they are expressions of the problem"* (Lloyd, 2021). We should invest our time not into managing behaviors, but into seeking to understand the context that surrounds them. We must let go of our desire to tell people what to do and lean into asking them what they need to be successful. We must embrace the fact that the behaviors we see are merely symptoms of much deeper problems within our work worlds – that they are "expressions of problems." We must understand that deep and profound insights and learnings can be gained by doing this, bringing us as close to a true sense of understanding as we will ever be. We must stop doing things to people – even if it feels really good and like the right thing to do – if we want things to actually improve.

Another quick word on psychological reactance

In a *Psychology Today* article written by Elizabeth Hall, Ph.D. (2019), she described reframing of experiences into ones that are not a threat to freedom – she highlighted this as one way we can avoid psychological reactance. Supporting her ideas, one study found that by telling participants that "they were free to decide for themselves what is good for them" after being told to do a specific health behavior, like flossing their teeth or wearing sunscreen, was able to reduce this psychological reactance (Bessarabova, Fink, & Turner, 2013) (Miller et al., 2007). The inclusion of choice, freedom, and involvement in the decision-making process, is vital to our approaches to the safety of work and Human and Organizational Performance – we are talking about this shift from doing things to people, towards doing things with people.

Reframing the experience

How can we reframe the experience of "safety" within our work worlds – moving it from something we 'do to' others, to something we 'do with others.' We must actively choose to do things with people. Before writing a new rule or procedure, how much input do you receive from those that it is intended to apply to? Before buying a new piece of equipment, do you spend time with those that operate it to better understand the daily challenges and complications they face? This questioning could extend on to nearly everything within your organization, so I will stop there and go a little broader – how much do you involve your employees in the decision-making process? Do you lean into their expertise? Do you listen? How about freedom and choice, do they have any choice or autonomy in the matter? It is a lot, I know! But this is the shift towards doing things with the people within your organization.

Here are some facts: you do not have to operate that fancy new piece of equipment that barely works, you do not have to spend 40 hours each week working with that horrific and time-consuming piece of software, you do not have to complete those endless checklists before every task, and you do not have to live daily with the problems all your "doing" has created.

A story about pre-job briefs...

I had the opportunity to witness the horrors of "doing to people," along with the shift towards "doing with people" recently – specifically, relating to the use of pre-job briefs. While working with an organization on their journey towards Human and Organizational Performance, one particular pain point that consistently emerged from their employees was the required completion of pre-task paperwork, more specifically the completion of pre-job briefs. I am certain that you know the piece of bureaucratic safety clutter I am referencing (sometimes referred to as a "take 5," task-safety analysis, etc.). This seemingly endless form had grown and grown over the years – typically with each passing event – into a gargantuan list of check boxes. Hearing these consistent frustrations piqued my curiosity; I just had to learn more. Over the course of several weeks, we conducted learning explorations and learning teams with various groups within the organization – speaking to many who were required to use the form along with those that controlled it – seeking to understand.

A few interesting learnings:

- Good pre-task conversations were happening, especially around jobs that were thought of as higher risk. The form was typically completed before or after the conversation, and then stuffed away in the supervisor's desk to show auditors (or other overseers) compliance
- From the employee's perspective, the form was more useful as a CYA (cover your ass), than as a tool to promote pre-task safety conversations – a common quote was that "we fill them out more than any other paperwork... after something bad happens."
- From the organization's perspective, the form was a vital safety tool, one that ensured these critical conversations were happening
- Those controlling the form felt that it was helpful to the end user, and helpful to preventing events

A key takeaway for the organization was that – with the best of intentions – they were doing safety to their employees. They simply had not felt the need to involve those required to use the form in the process of constructing it, of deciding what it should be, and how it should be used. So, the results were pretty typical of any other endeavor that involves a healthy dose of "doing stuff to people" – the process was wildly ineffective and a massive pain in the ass. It felt good, it looked good, it seemed like the right thing to do, it was trackable and measurable, and it was completely and utterly useless.

Experimenting to get better

Now what? Pre-job briefs, at least in the United States, are typically a regulatory driven program. Additionally, they often function as anxiety reducers for organizations and as mechanisms for individuals and organizations to display their carefulness and diligence around safety matters (Havinga, Shire, Rae, 2022). For many, parting ways with pre-job briefs is simply not an option. These are the circumstances that this particular organization found themselves with – a form that was not going anywhere anytime soon. They were stuck with a regulatory driven form, a form that the company was not willing to let go of, but one that was completely non-functioning and creating a massive amount of headache.

A general agreement was made with the leadership of this organization – one that allowed those that were the primary users of the process to create it. Only two rules were provided: 1) the process (and accompanying form) had to meet the minimums of the OSHA standard, and 2) It could no longer suck.

Working together in small groups, several teams consisting of end users of the process along with those that had historically controlled it, produced a handful of prototypes. These prototypes were then deployed for use within various field crews – each being asked to try out the new form and to provide suggested changes. Each team would regularly check in with the groups experimenting with their particular prototype to capture feedback and more learnings – using this input to make real-time changes to the prototype prior to redeploying it for further trialing. This cycle of testing and modification continued until each group felt that they had crafted the best product possible.

This is where things get just a bit more interesting. Keep in mind that these teams were experimenting completely independent of one another. After each group had a finished product in hand, they provided their form to another group for field trialing and feedback – they swapped work products. Now gaining feedback from other field crews and teams, the testing and modification process continued on for a few more cycles to polish their prototypes into final drafts.

These teams then met to debrief, share learnings, and to see if it would be possible to pull these various polished prototypes into a single form. From their continued conversations, and by continuing to test their prototypes with field crews, they did just that.

What did the final product look like? It was nowhere near what the organization had originally believed to be useful – a massive four-page form covering everything from shoelace tying to the control of hazardous energy. The new form – the one that was crafted with the people that had to use it – was simple, clean, elegant, and focused on the things that mattered to them. The areas the final product primarily focused on:

- STKY (shit that kills you)
- Lifesaving controls
- Verification of lifesaving controls
- Other STRM (shit that really matters)

The final form ended up being only one page – one page packed full of extremely meaningful things. The form went from being nothing but unused bloated safety clutter to, at the very least, something that the employees felt to be of value to them. This is the shift from doing things to people towards doing things with them – that is the power of doing things with our organizations.

People are the solution – ask them what they need…

KEY POINTS

Stop trying to tell people what to do – focus on asking them what they need

We must reframe the experience of "safety" within our work worlds – moving it from something we 'do to others,' to something we 'do with others'

Doing things with people is a deliberate act

Lean into employee involvement and micro-experimenting

PUTTING IT INTO
PRACTICE

- Seek to shift organizational assumptions that create a desire for the management, manipulation, and modification of employee behaviors as a mechanism of control
- Create opportunities to involve employees in the decision-making process
- Find areas for improvement and allow employees to micro-experiment and come up with their own solutions

10 IDEAS
TO MAKE
SAFETY
SUCK LESS

LEARN DELIBERATELY AND OFTEN FROM THOSE THAT GSD

CONVERSATION PRIMERS

How does your organization currently approach learning?

How often do you seek out an understanding of 'normal work?'

What is your organizations typical approach to learning from unexpected events?

10 IDEAS TO MAKE SAFETY SUCK LESS

"I was sick to my stomach – after realizing what had happened – I couldn't focus and had to leave work," a rather large and tough looking mechanic said while attempting to hold back his tears. "I could have killed someone…" he said, now wiping the flowing tears from his face. The mechanic continued to explain how he had locked out a piece of equipment – something he had done hundreds of times prior – in preparation for another crew's upcoming task. He detailed how he performed this lockout process methodically following the plan step by step, how he had to do it in the middle of the night due to the need to get this critical equipment fixed and back online, how he was unable to locate batteries for his flashlight, how it was verified by another employee as being "correct," and provided a wealth of other rich contextual information relating to the work.

"When we realized the lockout was wrong, I just broke down. It was like someone punched me in the gut," the mechanic said. He described his feelings of fear and disappointment, and his deep sense of responsibility for what had occurred. "I was afraid that I had just killed someone…" he again stated now gazing down at the floor. "What happened?" I asked. The mechanic explained that no one was injured or killed, that nothing was damaged, and how his initial fear that his actions might have killed one of his friends, now grew into a fear of losing his job. "I took a few days off to think," he said. "I just had to get away and think, I couldn't do my job because I was just too shaky…" he explained.

121

This employee was new to this particular company, having started with the organization a couple years prior to the event. It is also important to mention that, a few years prior to the mechanic beginning his career with the organization, they had started down the path of Human and Organizational Performance – embracing the use of learning teams specifically.

"I was really afraid that I was going to be fired," the mechanic said. He followed up this statement with "I should have been fired, I deserved to be fired..." I then asked – noting that he had not been terminated – how things had gone after the event. "It was so different," he said, now no longer weeping. "It was definitely not what I was expecting," he continued. The mechanic went on to explain how the company had embraced him, how they did not blame him, and how they involved him in the learning process. He detailed how he had not been interrogated, that instead, he was asked to be a part of a learning team for the event. "They asked me if I would be willing to be a part of a learning team, something I had never heard of before, to help make things better," the mechanic explained.

This story, one of a focus on restoration and learning, came from a recent experience I had working with a particular organization. We were conducting learning explorations into "how things were going" as the company continued its shift towards Human and Organizational

Performance – seeking out the good, the bad, and the ugly to better understand the present state of the organization after five years of a purposeful shift towards HOP. Several additional stories, like the one mentioned above, came to the surface during these conversational explorations into lived reality. Stories about how things felt different, how things were getting better, and of areas that could use focus.

What a great example of learning on purpose, of moving beyond blame in pursuit of learning, and of embracing learning from those that GSD (get shit done). The story above demonstrates just how much can change within an organization through the application of Human and Organizational Performance – this company could have easily been described as "blame and shame" focused in the years prior to their shift. What a difference in lived experience for the employees working within the company – raw and real conversations (like the one above) would have never taken place prior to the organizations move away from blame and shame. This shift towards Human and Organizational Performance (and the use of learning teams) allowed the involved employee to share their story, to share the "real-deal" story of their work, and to actively participate in learning – it allowed for the organization to gain raw and real information. The result? The involved employee, along with the organization, had a positive and meaningful experience that resulted in a wealth of operational learning.

Focused and less focused methods for learning

This story highlights the use of two particular methods of gaining vitally needed operational intelligence, the use of learning teams and the use of learning explorations. A learning team can be defined as a way of looking at safety, quality, and operational excellence differently – by involving and empowering those that do the work – to drive improvements at both the worker and organizational level (Sutton, McCarthy, Robinson, Conklin, 2020). Learning explorations can be described as more of a "freestyle" and organic approach – based upon the general concepts and principles of learning teams – that allows organizations to seek out (while learning a lot along the way) opportunities for deeper and more focused operational learning.

I tend to apply these methods loosely based on what we are looking at and based on the outcomes we are hoping to achieve. Often, learning teams are used to explore problems or betterment opportunities, seeking out valuable learning rich and contextualized information, in hopes of generating fixes for particular pain points, issues, or problems. Learning explorations should be viewed as more of a less granular starting point – the casting of a wide net – to discover opportunities that we should seek to learn more about. While some fixes are usually captured, formal problem resolution is not the end goal of a learning exploration. I typically lean towards the use of more structured and formal learning teams around things like events, operational surprises, the solving of specific pain points or problems, and other more specific examinations – while utilizing these even less rigid and less focused learning explorations to take a broader view of things like current organizational reality, lived experience, organizational stories and lore, effectiveness of overall approach, and other less specific areas of interest. Often, learning explorations result in the use of more focused and specific learning teams to solve problems or explore things deeper, based on the learnings gained during the explorations.

Some key operational learning principles and concepts

According to The Practice of Learning teams (Sutton et al. 2020), there are five core principles of learning teams:

- Seeking to understand work-as-imagined and work-as-done provides valuable contextual information
- Groups outperform individuals in problem identification and problem solving
- Workers have the best knowledge and understanding of the problems they face
- The more effort put into seeking to understand the problem, the better the solutions that emerge

125

- Providing "think" or "soak time" for reflection drives learning and improvement

We are seeking to learn from those that GSD (get shit done) because only they have true knowledge and experience of how things happen in real life – we are hoping to tap into reality. The people that do the work best understand the work. They know where things just do not work, where they must make do, where they must stretch, and where our systems underperform, create headache, or must be worked around. Since we simply cannot understand the reality of work, we must deliberately seek to learn from those that do. We seek out this vital operational intelligence because we understand that people are the solution, and that we need to ask them what they need to be successful (Dekker, 2014). Without this 'looking glass' into the reality of work within our organizations, we are practically operating blind.

We have an image of what work looks like, how work should go, how successes are created, and how events occur, engrained in our minds. The problem is that how we imagine things happening, and how things actually happen, are two vastly different things. The fact of the matter is that we will never, with 100% accuracy, understand how things get done within our work worlds. But we should seek to understand it better by allowing those that do understand (the people that do the work) to teach us the reality they face daily – we should seek this learning deliberately and frequently. The better we understand the realities of work, the better our work worlds can become.

Learning team basics

So, you want to conduct a learning team, now what? Before we examine the "how to" of performing learning teams, allow me to insert a warning here – something I have seen occur in many organizations. Do not overcomplicate the process or get too hung up on structure – a bit of the idea here is that these approaches are less formal, less rigid, less linear, and more open and real. Do not allow your organizational desires for uniformity and repeatability to stand between you and real learning. Avoid falling into the traps of over proceduralization, of seeking absolute control over the process, of killing learning and innovation via the application of strict rules. Avoid bastardizing this simple and immensely valuable tool into something that it is not – at all costs.

The most basic steps used to conduct a learning team can be described as 1) learn, 2) soak, and 3) solve. These steps are commonly expanded into five steps.

The five steps of a learning team:

1. Prepare
2. Learn
3. Soak
4. Improve
5. Act

Let's take some time to explore each of these steps in greater detail...

Prepare

After identifying the need for a learning team, you will need to take some time to adequately prepare. During this time, you should gather any supplies that you think will be needed (I am pretty fond of flipcharts), along with compiling any required information (such as event information, technical data, manuals, etc).

Learn

This first meeting of your team should only be about learning as much as you possibly can. At this point in the process, we should be purposefully avoiding diving into "fix it" mode. This sessions time should be dedicated to the team discussing and discovering how work actually gets done.

Soak

This reflection point is one of the more vital parts of the learning team process (Sutton et al., 2020), do not skip it. As people, we need time to soak information up and to allow for ideas and other information to bubble up to the surface. In practice, I typically separate the first and second meetings by about a day. Too much time between meetings results in forgetting or losing track of vital information or ideas – placing the meetings too close together does not allow for enough time to think. This time between meetings can vary – focus on trying to find "the sweet spot."

Improve

In this second session, after a brief review of all that has been learned so far, you will begin to move the conversation towards improvement – taking all that you have learned from the first session and turning it into real ideas on how to render things better.

Act

Now we begin to turn these ideas into actions that add defenses, remove error inviting situations, fix problems or pain points, and make things better. Whatever method your organization normally uses to track and complete actions will typically work here, just do not allow a burdensome action tracking process to dissuade people from using or participating in the learning team process.

How these five steps play out...

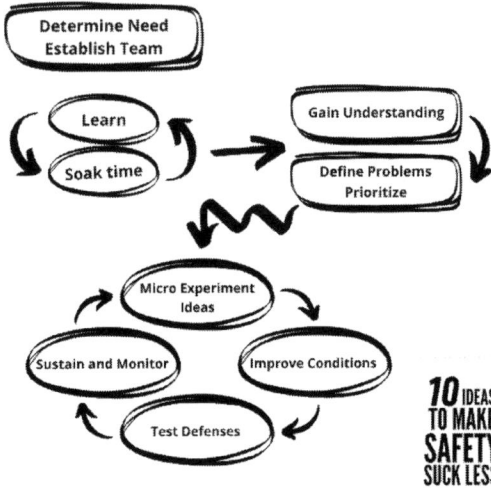

Learning Team Cycle
(adapted from Edwards, Baker, Conklin)

Determine Need
Establish Team

Learn
Soak time

Gain Understanding

Define Problems
Prioritize

Micro Experiment Ideas

Sustain and Monitor

Improve Conditions

Test Defenses

10 IDEAS
TO MAKE
SAFETY
SUCK LESS

When to use learning teams

Learning teams can be used anytime that you need to learn more about something. Things like events, interesting successes, and particular pain points, are all great areas to dig in deep and seek to gain a better understanding. Learning teams can practically be used anywhere, but we must acknowledge the realities of time, budget, and resources that we face within our organizations. A problem I often see is that, once the power of learning teams is recognized within an organization, there is a desire to do a learning team for nearly everything. A healthy dose of prioritization is key to not going "learning team crazy."

Another key item I have found helpful in the avoidance of going "learning team crazy" is the embracing of micro learning teams – the allowing of smaller and more organic worker led learning teams to solve crew or group problems. Many organizations dissuade these occurrences due to their independent nature – fearing that the learning team will not follow their process or that they will not share learnings throughout the organization. You should not only permit these micro learning teams, but you should also encourage them. Their independent nature – something most organizations cite as negative – is exactly what makes them highly valuable. Additionally, do not get so hung up on the sharing of learnings that you cause learning to not occur. The people that work within your work world will share information – information that they deem to be worthy of sharing – with others throughout your organization. I have witnessed it time after time, a group or crew of workers proudly sharing up through the organization how they have made things better. Will you know about every learning opportunity or learning team that occurs? No – you would not know that information even with the tightest grip of oversight and control. Do not panic, embrace some of this "not knowing," encourage your employees to go out and learn, and reap the rewards.

131

Let's take a look at a few areas in which the application of more formal learning teams would be appropriate:

Event learning teams

Keeping in mind that a learning team is an operational learning tool that brings those that are closest to the work together to describe how work is actually being accomplished in the field (Edwards and Baker) – a post-event learning team brings together (appropriate) stakeholders and employees connected to the event, to seek to learn the story of how each person saw the event, the story of complexity and normal variability, and the story of normal work, to improve our understanding of processes and systems (Sutton et al., 2020).

Post-event learning teams should not:

- Look for blame
- Act as investigations
- Focus on monocausality

According to Edwards and Baker, our goal must be to learn enough that we can understand the perspectives of those we are learning from – that we could easily see ourselves in their shoes. By understanding the conditions the involved employees were presented with, the information and local indications they had, the tools and equipment they were using, and the pressures they were under, industrial empathy is created (Edwards et al.).

Interesting successes

As organizations, we spend a lot of time focused on trying to understand how things go wrong, investing vast amounts of time and resources into these pursuits. We seem to be far less curious about, or concerned with, the successful work occurring within our work worlds every day.

We typically draw a broad assumption that negative consequences are the result of negative causes – that if we are not experiencing negative events then nothing negative is occurring – if the outcomes we are experiencing are good, then surely all must be well (Dekker, 2014). We fall for this trap quite often. But our work worlds are wildly complex – even bad process can lead to good outcomes and good process can lead to bad outcomes (Dekker, 2014). Seeking to understand how normal work happens – the reasons why work is usually successful – is an area we should be obsessively curious about.

Keeping in mind that every organization, even the largest ones, have finite amounts of time and resources, we must apply some level of sorting and prioritization to our learning endeavors – hence the "interesting" bit. Focus in on things that are learning rich – work that went extremely well, work that should not have gone well but did anyways, and other interesting "bright spots" within normal work.

Pain points

Listen attentively for signals of pain within your systems and processes. Pain points within our work worlds, like physical pain occurring to us as people, is often a signal of danger – pain typically acts as a protection mechanism to steer us away from harm. Pain points, like other weak signals given off by our systems, are signals of trouble on the horizon. Listen for and lean into these pain points as they are often learning rich opportunities for organizational betterment. Pain points often sound like:

- That thing never works right...
- It's way too hard to...
- I don't know why we...
- It's so dumb that we have to...
- We must make do with...
- We can't get...
- And many more...

Pain points contain a wealth of information about how work normally happens – typically telling the story of workers overperforming and making things happen in a system fighting against them. They are opportunities we should not ignore. Pain points are starting points for deep and meaningful explorations into the challenges workers face daily – they can lead us to rich learnings and help us begin the process of rendering our systems better for those they are supposed to support.

The use of learning explorations

As previously mentioned, learning explorations can be described as a "freestyle," organic, and conversational approach used to seek out opportunities for deeper and more focused operational learning. When utilizing learning explorations, we are casting a wide net to see what can be hauled to the surface. These sessions are not necessarily focused on the generation of fixes – learning explorations are about listening for areas in which we should seek to learn more.

The four basic steps of a learning exploration:

1. Identify an area for exploration
2. Conduct sessions
3. Evaluate
4. Seek deeper learning

Let's take some time to explore each of these steps in greater detail...

Identify an area for exploration

In the example provided earlier, learning explorations were used to pulse company progress relating to the adoption of Human and Organizational Performance – seeking to better understand where things were going well, not so well, and areas for improvement. Learning explorations can be used in almost any situation where there is not an identified pain point or problem in need of fixing, but it is believed that pain points, problems, or other rich learning opportunities exist.

Go broad but avoid being overly broad in your approach – there must be some level of focus applied to the use of these explorations or you run the risk of information overload. Define the parameters of your exploratory efforts beforehand; what are you hoping to learn more about? What do you think could use some attention? Where have you heard whispers of potential learning opportunities? Map out your exploratory questions or "conversation starters" from there.

Rather than asking overly broad questions such as "Can you tell me how things are going?" ask things more along the lines of "We have been putting effort into bettering (insert something here). Can you tell me about your experiences with that?" or "Can you teach me about your experiences with (insert something here)?"

Conduct sessions

Conducting a learning exploration is very similar to conducting a learning team – one notable exception being the absence of a second session and soak time. Remember, we are not seeking to solve problems, we are seeking to learn about problems that we do not yet have knowledge of. These learning explorations are more listening and triage, than deep dives and problem solving.

A learning exploration can usually be completed in a single session – averaging roughly 90 minutes in my experience. But learning explorations typically consist of several onetime sessions conducted across several different groups – think different groups or crews that share or work within the same systems. Again, we are aiming to go a bit broad here and capture a wide swath of contextual information.

Evaluate

With this new information now in hand, it is time to comb through it all. Various bits and pieces can (and often should) be lumped together into larger overarching categories. As an example, during some recent learning explorations with an organization, various problems with engineering drawings kept surfacing during our sessions. Each concern was a little different – from missing drawings to the inability to print drawings – but it was a very clear indication that there were problems deeper in the system. Each of these unique pain points were categorized as "engineering drawings." This information was then used to go out and conduct focused learning teams to generate specific system improvements.

Additionally, some things that you unearth will be standalone items. These can sometimes be

quickly moved into more formal learning teams –
some might even be direct fixes. Whatever the
case, be sure to evaluate this information
thoroughly, and be on the lookout for
opportunities to either directly better the issues at
hand or dig deeper into them.

Seek deeper learning

Now, with sorted and prioritized areas for deeper
learning, it is time to do just that. This is where
we can begin to do follow ups, fix specific issues,
or begin more formal learning teams to work on
the pain points, problems, or betterment
opportunities discovered during our learning
explorations.

Some basic conversation starters for learning explorations:

- How are things going with (insert something here)?
- With (insert something here), what are we missing?
- We started doing (insert something here), how is it working for you?
- It seems like we are really good at (insert something here), why do you think that is?
- We have gone through a lot of change lately with (insert something here), how has your experience been with that?
- Can you teach me about your experiences with (insert something here)?
- And many more...

Getting started

I come across many organizations that are paralyzed by a fear of trying out something new or different. Do not be fearful of giving learning teams or learning explorations a shot – try one on for size and see how it works. I can promise you that you will not be disappointed by the outcome. I would encourage you to start small and feel your way through the process – give yourself room to fail in a small-scale setting.

I often hear things from companies like "we can't do that, we are required to do a root cause," or "our current procedures require us to do X, Y, and Z," using these rules and requirements as roadblocks to the use of learning teams. I understand these challenges well, having lived through them myself while working directly for organizations trying to start on their Human and Organizational Performance journeys. How did we work through these barriers to implementing the use of learning teams? We did them anyways – we really did just go out and do them. While your organization might have a rule that requires a root cause analysis to be performed, it is highly unlikely that it has a rule that bans the use of learning teams. Around events in particular, we started doing both – meeting whatever the procedure or rule required, while also performing an independent learning team as well. Doing double the work is never fun, but as frustrating as it was, it was well worth the effort. Doing both allowed us to easily demonstrate the difference between the two methods and compare the amount of learnings gained.

If that is still just a little too much for your organization, lower the barrier to entry that much more and focus on simply trying to learn about areas for improvement. Pick out something that people are struggling with, an area of headache, or an opportunity for betterment, and go out and give it a try. Find a problem that people are facing and use a learning team to solve it. Do this a handful of times, tell the story of this rich and contextual information up through your organization well, and the progress of these efforts will not be ignored. Often the only true barrier to the use of learning teams, is simply the act of starting to do them.

Be curious, seek to understand, and go out and learn deliberately and often from those that get shit done.

KEY POINTS

Use learning teams or learning explorations for anything you would like to learn more about

Pay close attention to learning about normal work – seek to tap into lived reality

Be cautious to not create too much structure or rigidity around the process

Do not be afraid to get started – start small and experiment, then go big

PUTTING IT INTO
PRACTICE

- Deliberately seek out learning rich opportunities to get started
- Start small, and conduct a learning team in a 'safe to fail' setting
- Begin by experimenting and feeling your way through the process
- Focus on sharing these context-rich conversations up through your organization

10 IDEAS TO MAKE SAFETY SUCK LESS

PAIN POINTS ARE STARTING POINTS

CONVERSATION PRIMERS

Does your company currently invest time into discovering and learning about organizational 'pain points?'

If yes, how so?

What is the organizations' normal reaction to employees bringing up 'pain points' or issues?

10 IDEAS TO MAKE SAFETY SUCK LESS

What is pain exactly? Beyond that achy back or tooth, that horrible jolt received after grabbing a hot pan from the oven, or that horrific sensation felt when stepping on a child's toy in the middle of the night – What is pain, and what is its purpose?

The purpose of pain

Pain primarily functions as a defense mechanism, one designed to steer us away from harm. Pain provokes an unconscious physical response, and it is there to warn an organism that something is causing them damage and that they should do something about it – like remove our hand from that hot pan or remove our foot from that toy (Munro, 2015).

The importance of pain is pretty straightforward in this situation–we might not realize that we were touching that hot pan – pain forces us to let go, helping to prevent irreversible damage. The same goes for stepping on that toy, upon feeling that sharp stinging sensation, we shift pressure to the other foot to minimize damage. While it is easy for us to dream about a life absent of pain – no more headaches, stinging sunburns, or aching backs – pain is vital to our survival.

The purpose of pain in our work worlds

The purpose of pain points within our work worlds is not all that much different than why our bodies feel pain – pain is a signal that something is wrong, something is not working, and that there is a high likelihood of greater trouble on the horizon if we do not take action – something is causing us damage, and something must be done about it. Like those stinging sensations experienced from grabbing that hot pan or stepping on that *LEGO* while in search of a midnight snack, our organizational pain points are trying to tell us something as well. Pain is often telling us that we need to act, that we need to fix, or that we need to stop doing whatever it is that is causing us pain.

While we experience pain as individuals directly from things like getting a sunburn or filleting ourselves open with a razor knife (I have a nice scar from that one), our organizational pain points often manifest as struggles, challenges, pressures, and the like. These pain points routinely present themselves as things like a piece of equipment that constantly breaks down, a process or rule that makes accomplishing work nearly impossible, understaffed jobs or projects, rules that do not make sense, the inability to obtain needed resources or supplies, and various other areas that create headache and grief for those trying to accomplish work.

Common organizational pain points:

- Things that are harder than they should be
- People can't get what they need – tools, equipment, funding, help, etc.
- Frivolous rules and hard to follow policies
- Impossible to use procedures or guidance
- And many more…

As mentioned previously, pain points often sound like:

- That thing never works right…
- It's way too hard to…
- I don't know why we…
- It's so dumb that we have to…
- We must make do with…
- We can't get…
- And many more…

These pain points are often organizationally induced sources of annoyance and frustration for those trying to do the work – these pain generating problems have usually been created by our own hand. In most cases, we created these pain sources with the best of intentions. We had high hopes of rendering our workplaces a bit safer, more productive, more cost efficient, or a little "better" in some other way. We rolled out a new "something" or changed an old "something," with little to no meaningful input from those that actually have to use or live with that "something," and from our vantage point our "something" looks like a raving success. While we pat ourselves on the back for a job well done, our employees are forced to painfully adapt, create workarounds, make do, and figure things out, working within the new mess of problems we have created. We simply lack the right perspective, and we rarely seek it out.

Power to Influence
(adapted from Conklin)

One of the most notable differences between occurrences of personal pain and the pain that occurs within our work worlds is that we, as organizations (those up and away from where the actual work happens), do not experience these stinging and jabbing sensations directly. Unlike when we stub or toe on the couch or catch our shin on a drop hitch, those up and away in our organizations do not feel these pain sources directly. Those farther away from the coalface (those often wielding the most amount of power and resources to alleviate or minimize sources of organizational pain) cannot directly feel the pains felt by those at the coalface (those typically wielding the least amount of power and resources to alleviate or minimize sources of organizational pain). The CEO does not feel the pain of having to operate some new wonky and unreliable piece of equipment – they only see it as a cost savings. The safety director that issued a rule requiring the use of foam-lined safety glasses does not experience the pain of them constantly fogging up – they only see a greater level of protection. The manager or planner that understaffed a project does not have to work extra hours to ensure the job is completed on schedule – they only see successful completion of a project on time and under budget.

Vantage Points
(adapted from Conklin)

Looking Down...

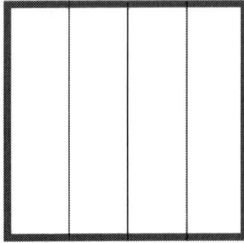

from above in the organization

Looking in...

from the coalface

10 IDEAS
TO MAKE
SAFETY
SUCK LESS

The pain sources within our companies that our workers face, whether they be organizationally induced, the result of outside pressures or influences, or born out of other complexities or complications of our industries or trades, must be acknowledged and deeply explored. To begin to do so, we must accept that we – those up through the organization – have a different vantage point, a different version of reality, than those getting the actual work done. By accepting this fact, and by leaning into the idea that we simply cannot understand the reality faced by those with their boots on the ground, we should begin embracing an extreme operational curiosity – a constant desire to seek to understand more about how work normally occurs. Learning more and more about normal work will allow us to uncover, and work with those that do the work, to better or remedy these points of frustration and grief.

Our typical reactions to organizational pain

One of *the 5 Principles of Human and Organizational Performance* teaches that "*how you respond to failure matters*" (Conklin, 2019). When faced with anything unexpected, especially things that are negative or unwanted, we often react with emotion rather than respond with grace. Let's face it, some of what we learn within our work worlds is seriously not good, and sometimes it's downright scary. Upon receiving bad news, we allow our emotions – often fear – to get the better of us and we panic, freakout, lash out, ignore, or breakdown. We are frightened by the news we are hearing – our minds begin to swiftly react – the perception of this threat activates the sympathetic nervous system and triggers an acute stress response that prepares our bodies to fight or flee (Psychology Tools, 2022).

"We are more often frightened than hurt; and we suffer more from imagination than from reality."

Seneca

This stress response can result in our imaginations running wild, along with our reactions. We freakout learning about these things that scare us, challenge our personally held beliefs or views about the working environment or organization, or when learning of other gnarly bits in our work worlds – things that we would rather not exist. Let's take a look at some common reactions that occur when those in a position of authority hear about organizational pain points or problems:

Deny

One typical reaction leaders demonstrate upon hearing about pain points experienced by their employees is to actively deny that the pain sources exist. Denial is a defense mechanism in which a person refuses to recognize or acknowledge objective facts or experiences. It's a process that serves to protect the person from discomfort or anxiety (Denial, n.d.). For some leaders, hearing about these pain sources generates far too much discomfort for them to stand. This discomfort or anxiety drives them

158

towards denial – denying that the problem exists, that it ever existed in the first place, or leading them to believe the problem to be "overblown" or "not a big deal."

Complaining!

Sometimes upon learning about the not-so-great things employees are experiencing, leaders gravitate towards chocking these areas of frustrations up to gripes or complaints. This can be heard in reactions such as "You know how John is…" or "This again?" or "Just do your job!"

Can't fix it!

We'll often hear this response manifest as things like, "Well, what do you want me to do about it?" or "I don't have any control over that!" or "I can't fix that!" Rather than seeking to learn more about the issue, rather than working towards bettering the problem or looking for solutions, this reaction is admitting defeat – choosing to not invest energy into problem solving – and accepting things just how they are.

Lashing out or blaming

Hearing about pain sources can sometimes cause leaders to lash out at or blame those bringing attention to an issue. Lashing out can happen for multiple reasons such as self-protection, past trauma, seeking to devalue and control, or stress (Borschel, 2021).

Typically, upon learning of these painful negatives from their employees, leaders will react with a mixture of the above reactions or defense mechanisms. Leaders will deny that a problem truly exists because they believe their employees concerns to actually be complaints. Believing that these pain points are simply complaints – that they do not exist or that they are not a real problem – the leader will gravitate towards "can't fix it!" mode, and then towards lashing out at the employees for their "complaining."

The problems with our typical organizational reactions

The most obvious issue with our typical reactions to learning about problems or pain points is that things do not get fixed. We have been gifted valuable insight into the problems faced by workers – things usually in dire need of fixing – and we choose to not render things better. An even bigger issue is that, by not fixing things, these issues tend to get much, much worse. Like that achy tooth that overtime turns from bad to worse, these pain generating problems faced by those accomplishing work rarely remain static – they are ever degrading and becoming more problematic the longer they are ignored or avoided.

These types of reactions leave employees feeling ignored. Often, people attribute being ignored to a belief that they are not significant enough to warrant any attention (Williams, 2009), signaling to our workforces that they, along with their pain points, problems, and challenges, are not worthy of the organization's attention. Employees find themselves left with a bag full of problems – issues with no solution in sight – and an employer that appears to not care.

The ways in which we react to employees bringing forth these pain points (or practically anything else for that matter) either encourages or discourages them from bringing up other issues. A poor reaction signals to the workforce that one should not bring up problems, because it is simply not safe to do so. Poor reactions to problems or issues from the organization drives employees towards silence – poor reactions cause companies to grow quieter and quieter over time. There is no amount of "raw" and "real" truth that should scare us as organizations. But silence? Silence should scare the shit out of us.

As Todd Conklin has so famously said many times before, "knowing less does not make you smarter." When we react poorly to "bad" news, when we decide to deny or ignore problems, when we chock up the concerns of the workforce to "complaints," when we decide that we just can't fix things or render things better, or we lash out at or blame our people, we are opting for silence – we are choosing to know less.

Pain points are starting points…

"Where there is pain, there is growth…" Pain points are starting points as they often lead you towards rot that is buried deep beneath the surface. The operational pains that we see manifesting within our work worlds – these signals of deeper problems in need of attention – they are windows into the challenges and struggles of normal everyday work. Pain points are usually symptoms of much deeper issues buried within the complex sociotechnical systems that make up our organizations – they are starting points for deeper exploration and learning. These whispers of "we have to make do…" "I don't know why we…" and on and on, are learning rich opportunities we must not ignore or avoid. We must lean into these pain points, we must seek them out, and we must see to it that we learn deliberately, deeply, and often from those that experience them.

A few of my favorite questions for examining for pain points:

- What is harder than it should be?
- What is the toughest part of your daily job?
- What is the dumbest thing you have to do working here?

The use of learning explorations or learning teams is a phenomenal way to learn about the existence of, or more about, the pain points employees face in everyday normal work. But even a simple conversation goes a long way in discovering the existence of these headache and heartache generating difficulties faced by those getting shit done. As I wander from company to company, from location to location, one of my favorite things to ask is, "what is the dumbest thing you have to do working here?" Sure, it's a little provocative, it's most certainly not "corporate approved," and it definitely earns me some strange looks while I'm waiting on my order at *Starbucks*, but it's a deeply meaningful and powerful question. The thing is this, every person I ask this question has a near immediate response – those doing the work are typically quick to share their struggles, hardships, problems, and pain points. People are more than willing to give you the "raw and real" facts about their jobs; people want to tell their stories. Often, they are just waiting for someone to ask – they are

163

waiting for someone to demonstrate true curiosity about what it is they do, and the challenges they face while doing it.

Pain points and various other problems become painfully obvious after something bad happens – after something breaks down, something catches on fire, something explodes, or someone gets hurt. It is easy to look back and see these now "loud" signals of trouble that have grown into seriously not good outcomes. Do not wait for these faint whispers of failure-in-motion – indicators that things have not "gone wrong," but that they are "going wrong" – to grow into loud and glaring failures. These pain points and struggles will never present themselves to you on a silver platter, you must go out and actively seek to discover them. A simple conversation, one spent exploring something like, "what is the toughest part of your daily job?" is invaluable. If you want to know where things are painful, where things are nearing failure, where things are going wrong, a good idea is to simply go out and ask those nearest to the work.

How pain plays out in our work worlds

The thing about pain points is that they are rarely monocausal. Once you have uncovered a particular pain point – let's say a new piece of equipment that is resulting in headache for those that are required to operate it – you will often discover that the source of pain has been born from many causes or for a variety of reasons, and that it often results in other new sources of pain. At the coalface or symptomatic level, we will hear something like "this piece of equipment is constantly breaking down." But as we dig deeper beneath the surface, we discover that this new piece of equipment was selected because it was 20% cheaper than what was used previously, that it was deemed to be "safer" than other models, that engineering believed that this new equipment would create less headache for those operating it, and that industry peers reported great productivity gains by using this new equipment.

Everything on the surface pointed to this being a good decision for the company to make. As we continue to dig beneath the symptom of "this piece of equipment is constantly breaking down," we discover a wide array of other issues. We find things like a lack of correct tooling or parts, employees being forced to put themselves in harm's way to keep up with production demands, equipment being ran beyond its operational limits to make up for downtime, and some other seriously scary stuff.

Now, let's complicate things just a little bit more...

This new piece of equipment is indeed frequently breaking down – at least once or twice per week – but parts are cheap and readily available, supply chain specialists have ensured that tooling and parts are always on site and ready to go, and those responsible for operating this equipment have learned how to fix it and fix it fast. In addition to learning how to get the equipment back online quickly, operators have learned how to push the equipment beyond its specifications to increase output. This higher output allows for the make-up of production losses during downtime and keeps production measures looking good (measures that are tied to incentive bonuses, ones that must always be met or else). From above, and without any knowledge relating to the reality of the situation on the ground, all looks well. Month after month, all that is seen up through the organization are high production numbers and operational goals being met and often exceeded.

Now, let's add a dash of some typical company reactions to pain...

Those living with these pain sources, those responsible for operating the equipment and keeping it running, have brought up the problems several times. Time after time, their concerns seem to have fallen on deaf ears. Local leaders scoff at concerns brought up about the new equipment, they quickly highlight how much

166

better (safer, cheaper, more efficient) they believe the equipment to be. Managers point to the positive metrics as proof of things "going well" and use them as ammunition for shooting down concerns. Corporate leaders, in the off chance that these concerns make it to their level, call attention to the cost savings generated, the engineering evaluations conducted, and the safety studies performed, as evidence that these concerns are non-issues. The concerns of those actually doing the work are denied, ignored, or chocked up to be frivolous complaints – the organization has purposefully decided to know less.

If we are not proactively listening for, and seeking to resolve these operational pain points, what actually happens over time? Workers continue to adapt and make do – band-aiding these problems and skillfully adapting through these issues the organization has presented them with – making sure things get done and work goes well. From above, all that is observed is successful outcomes, or work "going well." These systems continue to chug right along – with an extreme dose of TLC from the workforce – until suddenly they do not. The equipment once again breaks down, but the employee that knows how to repair the broken part resigned last week. The piece of equipment once again starts malfunctioning, but this time replacement parts are stuck in some warehouse far, far away. The equipment again trips offline, but this time a mechanic falls to their death while climbing up to manually restart it. All seemed to be going great – metrics were being met and money was being saved – until suddenly and violently everything catastrophically collapsed.

Sadly, this is such a common story in our work worlds. It is a tale of weak signals, of failure-in-motion, and of meaningful indicators that are unrecognized or ignored. It is a story of employees pointing out pain points and frustrations, only to be unheard by those above them. It is a story of focusing on targets and metrics rather than listening to the murmurs of system brittleness that, in this case, eventually grew into catastrophe.

Can you imagine the difference a little operational curiosity would have made here? While no one can say whether or not the failure could have been completely averted, the deliberate act of learning more would have at least stacked the cards in our favor. Imagine that the pain points experienced by these particular employees were explored deeper, that these problems and issues were discovered while their signals were still relatively weak, what would have changed? The answer, in my humble opinion, is everything.

Windows into deeper problems…

There is always a deeper story. The exploration of that deeper story, the pursuit of context, the seeking out of the "raw and real" stories of normal work, only serves to make our organizations better. When we encounter pain points in our work worlds, we must deliberately seek to learn more about them. These symptomatic pains are signals of trouble on the horizon, they are an indication that things are "going wrong," and ultimately, they are gifts, as they afford us an opportunity to discover "failure-in-motion" and respond.

Pain points are windows into deeper organizational issues and problems – they often lead you towards larger issues that are buried beneath the surface. We must be willing to pick away at these problematic scabs so that we can discover the festering infection that is so typically hiding beneath. We must acknowledge and set aside our desires to react, we must understand that we do not share the same reality as those doing the work, and we must embrace deliberate learning – learning is the only real tool that we have to make our work worlds better.

Pain points are starting points…

Dig deep.

KEY POINTS

Pain is a signal that something is wrong, something is not working, and that there is a high likelihood of greater trouble on the horizon

Pain points are often organizationally induced sources of annoyance and frustration for those trying to do the work

Pain points are starting points for deeper exploration and learning

Learning explorations and learning teams are a great way to learn about the existence of, or more about, the pain points employees face in everyday normal work

PUTTING IT INTO
PRACTICE

- Actively listen for pain points within your organization
- Deliberately seek out pain points by asking better questions
- Conduct learning teams or learning explorations to discover and dig deep into organizational pain points

10 IDEAS
TO MAKE
SAFETY
SUCK LESS

BECOME OBSESSED WITH THE THINGS THAT (ACTUALLY) MATTER

CONVERSATION PRIMERS

What is your organization currently obsessed with relating to safety, quality, or environmental?

Do you currently prioritize or sort these efforts?

If so, how do you determine the meaningful from the meaningless?

10 IDEAS TO MAKE SAFETY SUCK LESS

What are we obsessed with in our current approaches to the safety of work?

The simple answer is everything. Unpacking of that overly vague answer of "everything," typically reveals an aggressive and unwavering focus on the wrong things – a revelation that our so-called safety management systems drive organizational efforts and behaviors to obsessive extremes around the meaningless. These insignificant focus areas can be easily observed – from requiring that one must always don multiple layers of personal protective equipment to simply venture out of a jobsite trailer, to the performance of hundreds of documented behavioral safety observation cards each month, to the completion of endless amounts of pre-job paperwork, it really looks like we are "doing safety" – we have built quite the grand illusion of safety in our work worlds.

We plaster the walls of our workplaces with safety signs and slogans, we force workers to jump through various safety hoops on a daily basis, we drive leaders to spend their days "coaching and correcting" employees on the trivial (an example I recently observed: leaders tasked with coaching employees that had forgotten to carry their safety handbooks into the field with them), and we pretend that it makes a difference. We are indeed "doing safety," a lot of it in fact. But are we actually doing anything?

One particular example of this ongoing focus on the things that simply do not matter, is the continued (and seemingly growing) organizational passion around the use of pre-task safety paperwork. Commonly referred to as a Pre-Job Brief (PJB) in the United States – and as a Personal Risk Assessment, Take 5, Job Hazard Analysis, etc. in various other places around the world – these employee or leader completed safety cards have been believed to be a key component of any effective safety program for as long as many of us care to remember. The use of PJB's is considered (by organizations at least) to be a vitally important part of any job, and as a crucial tool for preventing unwanted outcomes. Companies spend what seems like endless amounts of time and energy monitoring it's use, and tallying completed forms – the thought being that, by completing these safety cards prior to doing work, employees will better plan safety into their tasks and will increase their awareness of the particular hazards they might encounter during the course of their work.

The problem? Pre-Job Briefs simply do not work. At the very least, PJB's do not do what we hoped they would do – i.e., better plan safety into work tasks, increase employee awareness, and, as a result of these things, prevent unwanted safety events. Workers have been telling us this for years in high-risk industries – think back to our conversation on Pre-Job Briefs earlier in this book – but we have chosen to not listen, believing that the use of these forms were "for their own good" and that surely, they must work, if only people cared more about their use. Recent research also indicates that there is no evidence that these cards are effective in reducing the risk of workplace accidents, and that they are more than likely just another example of "safety clutter" (Havinga, Shire, Rae, 2022). Yet, even with every shred of evidence indicating that we should lessen (or eliminate) our focus on these safety cards, their use persists in most high-risk industries.

Things look safe, so surely, they must be...

We have become fixated on the appearance of safety in our work worlds – believing that if we see "safety things," it means that we are doing "safety things," and that by doing "safety things," we are rendering ourselves or others "safe." But, like our beloved Pre-Job Briefs, so many of these "safety things" have little to no meaningful impact on the actual safety of work. These artifacts of our common safety management systems and strategies often only serve as highly visible "feel goods," as opportunities to demonstrate just how seriously our organizations take safety, as hard evidence of our due diligence relating to safety matters, or as – in the already mentioned case of Pre-Job Briefs as an example – organizational rituals that have the primary function of containing or minimizing anxiety (Havinga et al. 2022).

These "feel goods" and "look goods" are not benign. While these pieces of safety clutter may not serve to positively influence the safety of work as we had hoped, they do have an effect – our "safety things" often have harsh unintended consequences.

Unintended consequences…

Oh, the cluttered mess we have made… Now, our mess has been created with the best of intentions – typically in hopes of rendering our workplaces a bit safer – but it has resulted in a wide array of unintended consequences. In addition to the obvious problems that are created – things like increased administrative burden, the creation of employee apathy relating to company safety endeavors, and general frustrations and headache relating to these "safety things"– there are much deeper issues with our fixation on the doing of "safety things" in our workplaces.

As an example, Risk Homeostasis Theory proposes that, for any activity we undertake, we accept a particular level of risk to our safety in order to gain from benefits associated with that activity (Wilde, 2014) – If we perceive that the level of risk is less than acceptable, then we will often modify our behavior to increase risk exposure – if we perceive the risk at a higher than acceptable level, we will compensate by exercising greater caution (SafetyRisk, 2017).

Risk Compensation is yet another potential unintended consequence of all of this "safety work." With Risk Compensation our efforts to protect workers can backfire, resulting in smaller effects than expected or no effect at all, or even negative effects. Sometimes the risk is transferred to a different group of people, or a behavior modification creates new risks altogether.

These unintended side effects have been demonstrated in multiple pieces of research and literature such as:

- Peltzman, Sam. "The Effects of Automobile Safety Regulation." Journal of Political Economy, vol. 83, no. 4, 1975, pp. 677–725. JSTOR, http://www.jstor.org/stable/1830396. Accessed 24 Jul. 2022.
- Zolli, A. (2013). Resilience: Why Things Bounce Back (Reprint ed.). Simon & Schuster.
- Trimpop, R. M., & Wilde, G. J. S. (1994). Challenges to accident prevention: The issue of risk compensation behaviour. Groningen: STYX.

And many more...

There are numerous pieces of evidence to demonstrate these unintended consequences that grow out of our desire to render those around us "safe," with many of these examples coming from outside of the workplace. According to research published in *Psychological Science*, the wearing of a bicycle helmet might lead you to take more risks than you normally would because you feel safer while wearing that extra equipment (Gamble & Walker, 2016). A study of "childproof" safety caps on aspirin bottles found that the introduction of these "childproof" caps had no significant impact on reducing aspirin poisoning rates due to a general reduction in parental caution with respect to medicines – behavior adaptations such as parents leaving protective caps off bottles because they are difficult to open, or the increase of children's access to these bottles because they are supposedly "child proof" could potentially increase poisoning rates (Viscusi, 1984).

While concepts like Risk Homeostasis and Risk Compensation can be a bit controversial, the point remains the same – for every action we take, there are intended and unintended consequences. These ideas and examples illustrate just how important it is for us to be fully aware of the unintended consequences that can occur when we interact with complex sociotechnical systems – the safety of work included.

So, the creation and maintenance of this "appearance of safety" within our work worlds has had vast unintended consequences and could potentially be rendering them less safe. Additionally, our fixation with every minute detail relating to anything deemed "safety," seems to have left our work worlds lacking focus on what is truly important to the safety of work.

But…

We seek to manage and manipulate what we think we can – we touch what we can see. Many of the "safety things" we obsess over are highly visible, easily manipulated, and superficial items that can be best described as meaningless safety ritual. We can visibly see safety posters hanging on office walls, we can directly and easily influence the wearing of personal protective equipment by demanding that leaders harshly police this behavior, we can easily count the number of Pre-

Job Briefs and safety observations that have been completed, we can easily tally and track all of these numbers in safety spreadsheets, we actually witness those endless safety meetings and stand downs, and we can easily point to these highly visible items to highlight just how safe our work worlds appear to be.

But tragically, we have missed the mark. We have convinced ourselves that those check sheets are just as vital to protecting our employees' lives as the control of hazardous energy. We have fooled ourselves into thinking that if we can just develop some new measure – some hidden numeric treasure that will finally grant us predictive capacity – that it will work just as well as robust controls. We are certain that if we can just convince our people to be "more safe," then that is just as powerful as bettering the setting in which work takes place.

We have constructed this odd belief that says, we get better at safety by doing more "safety things." No matter the impact, no matter the efficacy, no matter the headache, and no matter the unintended consequences, we believe that the pursuit of more is how we better the safety of work. And more and more we have seemed to create within our work worlds, more headache, more heartache, more heartburn, more frustration, more distrust, more fear, and more pain and suffering – all while making little to no meaningful impact on worker safety.

Our focus on everything seems to have left us with a focus on nothing – a focus on nothing meaningful, at least. How can we move beyond this obsession with meaningless and ineffective "safety stuff?" By focusing on the things within our sphere of influence that actually matter.

An obsession with critical risk...

Let's cut right to the bone by saying this: You will not stop killing and maiming workers by focusing on the things that do not kill or maim workers. We have spent far too long overly focused on the things that hurt workers, while avoiding the things that kill them. But, as Conklin has so famously stated, "the things that hurt us are not the things that kill us..." With that in mind, let's begin to shift our safety obsession in a better direction by highlighting some basic areas of critical risk – better areas in which to place our operational efforts relating to the safety of work.

Three basic areas of critical risk most organizations face:

STKY – *Shit That Kills You*

These are the things that exist within our work worlds that actually kill or maim workers. Evidence suggests that serious injuries and fatalities are the result of some undesirable contact with energy (Hallowell, 2020). These energy sources include things like gravity, motion, electrical, chemical, and the like.

STRM – *Shit That Really Matters*

This can be nearly anything on the list of "things we must always get right." Think along the lines of environmental, quality, and reliability.

STBY – *Shit That Bankrupts You*

The Shit That Bankrupts you can include both STKY and STRM when left un or under controlled but can also easily expand into areas of regulatory concern, legal matters, company image, and on and on.

Impacting areas of critical risk

In his recent works, Todd Conklin has alluded to a 'sixth principle' of Human and Organizational

Performance – *controls save lives*. Nothing guards against critical risks quite like robust and error tolerant controls. These controls physically prevent energy from harming people, or they lessen the energy to a point in which the outcome or harm is minimal – think stitches versus amputation or broken leg versus dead, as examples. They stop an event, or they reduce the outcome of an event, even if everything else fails or there is an error or mistake.

Think of it like this, we should make it really hard to get seriously injured or killed – really hard to have catastrophic outcomes – and really easy to be safe – really easy to not have catastrophic outcomes.

Some basics of controls relating to critical risks:

Strong	Robust and non-brittle
Effective	Functional, Control > Hazard, sufficiently addresses critical risks
Error Tolerant	Easy to be safe, hard to be unsafe. Not operator dependent and functional even with the presence of human error
Verified	"I know because I looked, I know because I checked" Controls in place and verified effective
Periodically Tested	Defense testing, control assessments to examine for effectiveness – degrading, missing, non-tolerant controls

*Examining for the presence of lifesaving controls
or safeguards*

There are three primary questions – coined "*start
when safe*" by the Human and Organizational
Performance community of practice, and adapted
from the works of Conklin (Conklin, 2017) – that
are used to help reduce uncertain outcomes
relating to these areas of critical risk:

1. *What are the Critical Risks associated with
   the job (STKY, STRM, STBY)?*

2. *What Lifesaving Controls or Safeguards do
   we have in place?*

And...

3. *Are they enough?*

Let's take a look at how these questions play out in real life normal work…

What are the Critical Risks?

Today we are changing out a pump that has been malfunctioning. The STKY related to this task is the energy sources associated with the pump along with the lifting and rigging activities we will have to perform in order to swap the pump out.

What Lifesaving Controls or Safeguards do we have in place?

We have isolated the energy sources associated with the pump by locking them out. This lock out was independently verified and we have tested its effectiveness by attempting to start the pump, and by conducting a live-dead-live – everything checks out. The lifting area has been secured, we have verified that the rigging is well within its capacity to lift the pump, and we have a solid lifting plan.

Are they enough?

Yes, lifesaving controls are in place, verified, and tested. We have a strong lockout procedure and process that works well for us. We have a solid lifting plan and we have secured the area just in case something goes wrong with the lift, allowing for the load to fall within a secured area devoid of personnel. We are safe to start…

This extremely simple (but common) example demonstrates a crucial shift in focus towards the most important things – areas of critical risk – associated with this particular task. This is an overt focus on ensuring the presence of controls, and the building in of a little capacity or margin just in case they fail, to help to ensure that work goes well – even if it goes wrong. *"Start when safe"* is an acknowledgment of what is within our sphere of influence relating to the safety of work – that all we can truly manage is the presence of controls (Conklin, 2017).

Prioritization matters, and it matters a lot! Unfortunately, it really does seem as if we have completely avoided the subject – especially in the world of safety. We have continued to rally behind this idea that everything in safety is equally important, which has resulted in us only wasting time and creating more safety bureaucracy and headache, along with a vast array of unintended consequences. A healthy dose of reality and prioritization is needed to get us out of this hole, and to bring into focus what really matters to the safety of work. Without this needed and heavy-handed dose of prioritization, our organizations will continue to find themselves in the position of being safety junk peddlers – organizations that aggressively focus on the wrong things – pretending that we are making things better. We need to begin asking a simple yet powerful question as it relates to the safety of work: What actually matters? Especially as it relates to not killing or maiming workers.

Rather than a focus on everything, become obsessed with the things that (actually) matter... the things that actually kill or maim your employees.

KEY POINTS

Organizations have become obsessed with everything "safety" and lack focus on the things that are truly impactful

We have become fixated on the appearance of safety in our work worlds

We have constructed a belief that we get better at safety by doing more "safety things"

You will not stop killing and maiming workers by focusing on the things that do not kill or maim workers

Focus must be placed back on areas of critical risk, and nothing guards against critical risks quite like robust and error tolerant controls

Rather than a focus on everything, we must become obsessed with the things that matter, the things that actually kill or maim workers

PUTTING IT INTO
PRACTICE

- With a focus on areas of critical risk, begin to sort through and prioritize safety efforts
- Eliminate areas of meaningless safety work and begin to obsess over the meaningful
- Begin the use of "*start when safe*"
- Focus on managing the presence and viability of lifesaving controls and safeguards

10 IDEAS TO MAKE SAFETY SUCK LESS

MORE TOOLS

LESS RULES

CONVERSATION PRIMERS

How does your organization view the use of rules?

Do you currently use 'zero tolerance' policies within your company?

How do you determine what tools are needed by employees – how do you examine for usefulness of tools?

10 IDEAS TO MAKE SAFETY SUCK LESS

What if you were to wake up tomorrow only to discover that all of the top-down rules, the safety signs, the slogans, the pre-job checklists, and the procedures of your organization were no more? Poof! Gone! Vanished! Chaos and catastrophe would surely ensue... or would it?

A brief exploration into safety rules

Before we begin to explore this "what if," one that will be an "obvious" disaster for your organization – a company that now finds itself devoid of a rulebook – let's first start by defining what a rule is in the first place. For the purposes of this chapter, we will be focusing on rules in the sense of organizational or company rules that are prescribed to the workforce. *Merriam-Webster* defines a rule as *a prescribed guide for conduct or action. The Oxford English Dictionary* defines rules as *a set of explicit or understood regulations or principles governing conduct within a particular activity or sphere.*

A "Work rule" means a written regulation promulgated by the employer within its discretion which regulates the conduct of employees as it affects their employment (Law Insider, n.d.). If we view this definition through the lens of 'safety management' – creating the term "safe work rule" – this definition would become, a written safety regulation promulgated by the employer within its discretion which regulates the safe conduct of employees as it affects their employment. "Safe work" rules must always be followed – they are "thou shalts" – and are usually enforced through the application of heavy-handed disciplinary action policies.

Many organizations are in love with rules – especially safety rules – touting their "rules to live by," their "lifesaving rules," their safety handbooks, and their other endless lists of safety commandments. Many traditional safety management systems rely upon the creation, maintenance, and strict enforcement of safety rules as a crucial mechanism for influencing workplace safety. The strict following of "safe work rules" – as applied through more traditional approaches to safety – is believed to eliminate the probability of unwanted safety events. Many organizations persist in their beliefs that rules somehow create safety, but do they?

Do rules create safety?

Many traditional safety management systems place a heavy compliance burden on the end user of the system – the worker – through the expectation that all safety related rules be fully followed and complied with at all times and without fail. Practically any form of non-compliance – even those with a healthy dose of reason and logic – results in harsh disciplinary actions (like termination of employment) taken against those that have been found to be not strictly adhering to the rules. This belief grows from the idea that, through strict and unwavering rule following, accidents simply cannot happen. But what would happen if people truly complied with every single workplace health & safety rule applicable to their job? The short answer is nothing – work would grind to a screeching halt.

This fact has been demonstrated through the act of malicious compliance. Malicious compliance is the practice of following directions or orders in a literal way, observing them without variance, despite knowing that the outcome will not be what the organization initially desired (Staughton, 2022). Malicious compliance has been used a variety of times as forms of protest, and as a way for workers to strike. By rigidly complying with the rules – as they are instructed to do – workers bring operations to a halt or slow them down significantly.

When demanding rigid rule following – safety included – be careful what you wish for.

To pull this into our day-to-day lives for a second – imagine just how many laws and regulations apply to you at any given moment. How many can you recall? The most recent attempt at an official estimate conducted by the U.S. Justice Department – completed more than 35 years ago I might add – found that the federal government had defined more than 3,000 crimes in statute (Lehrer, 2019). It is pretty likely that you are always – even in this very moment – in violation of some law or regulation.

Let's do another quick experiment. How many rules does your organization currently have on the books? Let's take this just a bit deeper. How many "safe work rules" apply to you personally at any given time during the course of your day? Can you really say that you know all of your organizations "safe work rules," and that you follow them without fail? It's a pretty mindboggling thing to think about.

These traditional approaches – ones that are built upon the idea of strict worker compliance to create safety – rely on people to know (and understand) exactly the right rule to follow, to follow it at exactly the right time, and to do this every time without fail, in order to create successful outcomes. When workers ultimately experience some brush with failure, something like "non-compliance" or "rule violation" is quickly pointed to as the cause. Organizations then double down on rules and compliance, typically by adding even more rules and harsher penalties for non-compliance.

This "comply to be safe" ideology simply does not work. Our work worlds are dynamic and everchanging, even the most mundane or repetitive tasks are unique and new with each passing evolution. The circumstances and local variability of work-in-motion is constantly changing – changes that can never be known to those who draft rules or procedures – procedures or rules simply cannot guarantee safety because safety is not the result of rule following, it is the result of skillful adaptations on part of the worker while navigating this swirling mixture of pressures, resources, rules, procedures (Dekker, 2014).

A quick excerpt from 'Some myths about industrial safety' (Besnard & Hollnagel, 2012):

Humans constantly compensate for the discrepancies between procedures and reality and fill in the gaps between the procedures and actual operational conditions. This is the only way industrial operations can be conducted, given that it is impossible to anticipate all the possible configurations of a work situation and prescribe each and every step of an activity. It is human flexibility that compensates for brittleness of procedures, turning the latter into a useful tool for the control of systems and contributing to safety by doing so. Strict procedure compliance may even have detrimental consequences since it will limit the beneficial effects of human adaptation in response to underspecification of the work situation. Just think of situations where people 'work to rule.'

Denis Besnard & Erik Hollnagel, 2012

Simply put, the last thing that you should desire – especially if your hope is to render your work world safer – is rigid and 'unthinking' rule following. Strict compliance may be detrimental to safety and efficiency – Rules and procedures should be used intelligently (Besnard & Hollnagel, 2012). But for this to be possible, there must be a certain element of autonomy granted to the worker.

Back to our introductory question...

So, what would happen to your organization? Without all of the rules and procedures, all of the posters highlighting the need for strict following of rules and procedures, without all the safety handbooks and manuals, would your company actually devolve into chaos? No. People would continue to do what they have always done – get work done – safely and efficiently by adapting, figuring things out, and by creating safety on-the-fly.

Rules are an illusion of safety – they look great (literally) on paper. Rules really feel like they do a lot, they feel like the right thing to do, they feel like they should render workplaces safer, but they simply do not work very well. We have embraced the application of rules as a preferred method of control within our work worlds – large swaths of organizations that should be governed by norms and principles have been bureaucratized by rules – but rules are only an illusion of control. Understanding that rules do not actually create safety, should we burn all of these "safe work rules" from our organizations? Probably not, although a healthy dose of decluttering and rule elimination is likely in order. While reducing can be helpful, a shift in our thinking about rules is what is truly needed.

Common rules should naturally emerge from within your organization – as an agreed upon guide for acceptable behavior – rather than being imposed from the top down as a (perceived) mechanism of strict control. Shift your thinking away from organizational rules that must always be strictly adhered to, towards viewing them as organizational norms or principles that guide members of the organization towards commonly accepted actions. These "safe work rules" should become more of "safe rumble strips" – things that warn someone when they are drifting a little too far away from an already known and established path – rather than continuing to be used as rigid and enforceable "dos" and don'ts." Give people the autonomy to adapt and make decisions and seek to render them competent to make better decisions. Lean into their expertise and know-how and apply a more flexible and reasonable approach to these "safe work rules."

More tools less rules…

One of the biggest problems with these methods of organizational control – like work rules – is that they are just not very helpful to those accomplishing work. So much of this "shift in thinking" that we are talking about, is this idea of moving away from 'rules that seek to control,' towards 'tools that seek to help.'

A focus on tools that help…

Let's start by defining the term tool. *Merriam-Webster* defines a tool as *a handheld device that aids in accomplishing a task.* Now, we call a lot of things in our work worlds "tools" (and yes, sometimes even the boss), but are they really? Is that Pre-Job Brief really a tool? Is that booklet full of rules your employees must carry around forever and always really a tool? As organizations, we really like this idea of providing leaders and employees with "safety tools," but are they actually tools? Let's zoom in on a part of that definition, the part about "*aids in accomplishing a task."* For something to truly be a tool, it must be helpful – not hurt, slow down, or make things harder. A tool must actually help you accomplish something that is in need of accomplishing. It must be useful, it must be needed, and it must help one accomplish a task – if it does not, it is simply not a tool – it is likely just another piece of clutter. A valuable piece of the "tool" puzzle is to examine for the general helpfulness and usefulness in accomplishing work.

In various talks and presentations that I have delivered over the years, I typically provide what I have titled as "*The Home Depot* Example" to highlight this concept. To set the stage, imagine that you are neck deep in some renovations or a remodel, or that you are stuck fixing something that has broken at home.

You make your way down to *Home Depot* to pick up some needed supplies, and while venturing through the store and playing with all the neat things one might find in *Home Depot*, you come across a tool that looks like it might just solve whatever problem you are facing back at home. You are so excited about this tool that you don't even pay attention to the price – racing to get out of the store and back home to give it a shot. This new tool cuts your project time in half, it made the whole process a little easier, and made all that hard work suck just a bit less. That sounds like a pretty good tool to me.

Now, flip that example on its head. How likely would you have been to purchase this tool if it did not help you accomplish work in need of accomplishing, or if it didn't solve a problem you were faced with? Highly unlikely. What if that tool was advertised as being helpful, but upon using it you discovered that it just didn't work well? You would be searching for the receipt to return it to the store. What if it did the complete opposite, making the work slower, harder, and suck more? There is a high probability that you are going to throw it out the window, and then drive to *Home Depot* to curse whoever sold it to you.

Now compare that to the "tools" – especially the safety tools – we implement in our workplaces. How do they hold up – do they pass the "must be helpful" test? Many do not. With so many of our safety "tools" having stated goals like "to slow people down," "to make people stop," or "to ensure that people do (insert whatever here)," our safety "tools" are often not tools at all. Often, "safety tools" are just more mechanisms of employee control or "feel goods" that management can peer down upon as indicators of the presence of safety – safety "tools" are often just more safety work.

If a safety "tool" does not solve a particular pain point or problem, if it's hard to use and provides little (if any) benefit to the end user, if it makes work harder to accomplish (rather than easier to accomplish), or if it serves no purpose other than curing a bit of managerial anxiety around safety, then the intended users of the tool will bypass it, work around it, or simply not use it. If there is a rule or policy requiring the use of this particular tool, all of the above will still occur, but the end users will make it appear as if the tool is being used as to avoid being viewed as non-compliant by the organization.

If the use of a "tool" must be forced via the application of a rule, it is not a helpful tool. The best and most effective tools never require the use of force – people use them voluntarily because they are useful, and because they help solve some problem or challenge they face. If you have safety tool troubles – people not using them, avoiding them, working around them, or bypassing them – take a long hard look at the tool, not the person you want to be using it.

As we seek to provide employees and leaders with safety tools and resources, we must keep these concepts in mind. Whether you are examining preexisting safety tools within your organization, or seeking to create new ones, there are a few basic principles that always apply:

Some basic principles relating to effective tools

- It is needed – it solves a problem that is in need of solving, or aids in the accomplishing of a task
- It is useful to the person that must use it
- It is created with the people that need it

It is needed

Far too many safety "tools" are solutions in search of problems – they are broad non-specific "fixes" that don't actually fix anything in particular. There must be an actual need from the people doing the work – not from above – to warrant the creation or continued use of a tool.

It is useful

If a "tool" is not useful, if it does not aid in the accomplishing of work, or if it hinders efficiency, it simply will not be used. Reframing this as a question always seems to help, "Does this aid (insert person) in completing (insert task)?" Remember, to "aid" means to help, assist, or support someone in the achievement of something (Oxford Dictionary) – it does not mean to slow down, stop, or render inefficient.

It is created with the people that need it

Often, safety "tools" are created by safety people, rather than being created by those who actually need them. If your hope is to get the first two principles right, and your goal is to end up with an effective tool, lean into embracing the knowledge and know-how of the people that need it.

As organizations continue on their journeys towards Human and Organizational Performance, they often begin to see strict and rigid rules as being unhelpful, and as harmful in certain cases. As organizations continue to embed these H.O.P. concepts within their work worlds, they naturally begin to gravitate towards this idea of asking people what they need, rather than trying to tell them what to do.

A part of this journey is understanding and accepting that rules never create safety – no one has had their life saved by a 'life saving rule' yet – and that more rules do not render workplaces "more safe." Rather than seeking to add more and more rules – creating the illusion of safety – Human and Organizational Performance leads us to ask the better question of "what is needed to be successful."

Asking people what they need, and then supporting them with tools and resources that are based on their needs, helps them to be successful and does far more than rules ever will.

KEY POINTS

*Rules do not create safety, they create the
illusion of safety and control*

*Workers create safety in real-time by actively
adapting through the world that surrounds them*

*Strict rule adherence is likely detrimental to
safety and efficiency*

*Move away from 'rules that seek to control,' and
move towards 'tools that seek to help'*

*Tools must be needed, useful, and created with
the people that need them*

*Ask people what they need and support them
with tools and resources that are based on their
needs*

PUTTING IT INTO
PRACTICE

- Focus on shifting organizational assumptions around rules and rule following
- Let go of desires to blame and punish
- Eliminate unneeded or useless rules
- Build room for worker autonomy into your processes
- Focus on providing useful tools to those that need them – do this by deeply involving the end user in the creation process

10 IDEAS TO MAKE SAFETY SUCK LESS

STOP TRYING TO COMPLY (OR PUNISH) YOUR WAY TO EXCELLENCE

CONVERSATION PRIMERS

What level of significance does your company place on being in compliance?

How does your organization react to 'poor' audit findings?

When audits reveal areas of non-compliance, how likely is your company to seek out blame?

10 IDEAS TO MAKE SAFETY SUCK LESS

A compliant workplace is a safe workplace... or is it?

Let's take a moment to define the term "safety compliance," which, for the purposes of this chapter will exclude individual compliance – the act or process of an individual complying to a rule, desire, demand, proposal (Merriam-Webster) – and focus on the concept of organizational safety compliance. Safety compliance is often described as the act of adhering to safety rules set down by regulatory bodies and legislators and is an extreme focus area for many organizations around the globe.

Within our organizations, we have devoted massive amounts of time, energy, money, and resources to the pursuit of "safety compliance." While these efforts are somewhat well intended and focused on the creation and maintenance of a "safe working environment," there is an obvious self-serving or self-preserving element to these efforts. Through an organizations compliance with applicable regulations (or demonstrated due diligence attempting to comply with applicable regulations), they can often avoid or reduce costly citations, reduce exposure to potential litigation, and better maintain the public image of the company. A simple search of the internet reveals just how costly these infractions can be within the United States:

2022 OSHA Penalties

Type of Violation	Penalty
Serious, Other-Than-Serious, Posting Requirements	$14,502 per violation
Failure to Abate	$14,502 per day beyond the abatement date
Willful or Repeated	$145,027 per violation

Source: Occupational Safety and Health Administration – Penalties

With these risks in mind, and with a desire to render their work environments safer, companies painstakingly focus on safety compliance. They work diligently to complete a never-ending amount of detailed safety compliance assessments, they spend extravagant amounts of cash on external consultants to evaluate their workplaces against applicable safety and health rules and regulations, and they often create entire departments with their sole purpose being the auditing and coordinating of safety compliance.

The espoused aims of "safety compliance"

Safety compliance aims to keep workers, the public, property, and the natural environment safe from various work-related hazards through the adherence to safety standards and regulations. Basic compliance measures focus on the elimination of hazards from the workplace, the creation and maintenance of "compliant" internal work rules, procedures, and programs, the completion of compliance audits, and policy enforcement.

While the main common stated goal of "safety compliance" is to prevent workplace injuries, illnesses, and deaths, there is another element at play in this equation – one that is interconnected with this reduction in workplace injuries and fatalities. Safety compliance saves organizations cold hard cash. Safety compliance not only saves money around those costly citations and legal matters, but it is commonly proposed that – through the assumed preventative capacity of safety compliance – the costly burden of injuries and fatalities can be avoided. We see this demonstrated in how safety compliance is "sold" to organizations. Consultants and regulators are quick to point out that:

Total work injury costs for 2020 – $163.9 billion
Per worker – $1,100
Per Death – $1,310,000
Per medically consulted injury – $44,000

Source: National Safety Council – Work Injury Costs

These injury costs are pointed to as logical justification for increased effort and focus on safety compliance – reasoning that by seeking to become more compliant, or through the better maintenance of compliance in already compliant organizations, that catastrophic injuries and illnesses can be prevented. But is that true? Does safety compliance prevent events? Kind of.

A point of diminishing returns…

Compliance seems to have been beneficial in the reduction of occupational injuries and fatalities, at least until recently. In the United States this is evidenced by the steep decline in worker fatalities over time – worker deaths in America are down- on average, from about 38 worker deaths a day in 1970 to 15 a day in 2019, and the incidence rate of nonfatal injuries and illnesses among private industry workplaces occurred at a rate of 10.9 cases per 100 full-time equivalent workers in 1972 and 2.8 cases in 2018 (OSHA, n.d.) – since the introduction of The Occupational Safety and Health Administration (OSHA) in 1971.

Chart 1. Incidence rates of nonfatal occupational injuries and illnesses, private industry, 1972–2018

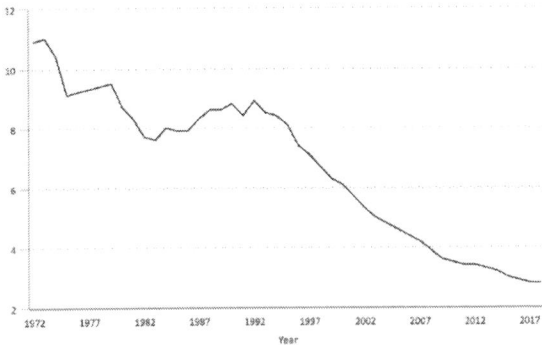

Source: U.S. Bureau of Labor Statistics, Census of Fatal Occupational Injuries.

A quick examination of the U.S. Bureau of Labor Statistics data surrounding the incidence rate of nonfatal injuries and illnesses among private industry workplaces shows two distinct points of

228

reduction relating to overall employee injuries and illnesses, with the first major reduction occurring directly after the formation of The Occupational Safety and Health Administration (OSHA) in 1971, and the other beginning in the early 1990's. This second more sweeping decline is commonly attributed to an increased focus on human performance and "systems" thinking beginning in the late 1980's (Dekker, 2019). A much more interesting observation is that, between 1992 and 2017, there was a sustained decline in overall nonfatal injuries and illnesses, yet occupational fatalities remained surprisingly consistent.

Fatal Occupational Injuries 1992-2018

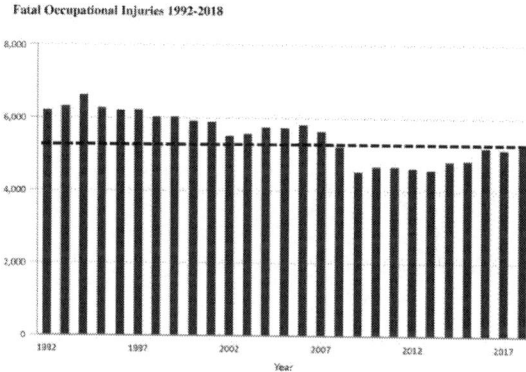

Source: U.S. Bureau of Labor Statistics, Census of Fatal Occupational Injuries.

Here is a detail of this data in case you are interested...

Year	Number of work-related fatalities
1992	6,217
1993	6,331
1994	6,632
1995	6,275
1996	6,202
1997	6,238
1998	6,055
1999	6,054
2000	5,920
2001	5,915
2002	5,534
2003	5,575
2004	5,764
2005	5,734
2006	5,840
2007	5,657
2008	5,214
2009	4,551
2010	4,690
2011	4,693
2012	4,628
2013	4,585
2014	4,821
2015	4,836
2016	5,190
2017	5,147
2018	5,250
Source: U.S. Bureau of Labor Statistics, Census of Fatal Occupational Injuries.	

So, where do we find ourselves currently? We find ourselves at a point of diminishing returns relating to our traditional approaches to the safety of work. Within our work worlds we continue to squeeze all that we can squeeze from these compliance efforts – and we have reduced many injuries and fatalities along the way – but we are no longer experiencing the same sweeping injury and fatality reductions we once did. No matter how hard we comply, we just cannot seem to comply our way out of killing people.

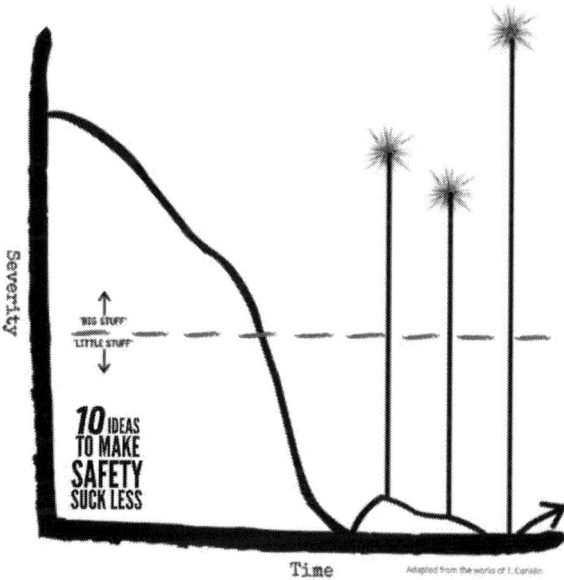

Safety compliance, along with all the other typical focus areas of more traditional approaches to safety, has resulted in the minimization (and sometimes elimination) of low outcome minor injuries within our work worlds – but fatal events are still happening at alarmingly consistent rates.

What gives?

An overfocus on "safety compliance"

Organizational desires to comply are typically born out of desires to prevent unwanted safety outcomes. It is a generally held belief within many organizations that by rendering their workplaces compliant with all applicable safety and health regulations, that they are rendering them "safe" – free from the potential risks of injuries and fatalities.

Select problems with an overfocus on safety compliance:

- "Safety focused" organizations have maxed out compliance
- Catastrophic events still occur in compliant organizations
- Safety becomes a bureaucratic activity to be demonstrated "up" (Dekker, 2019)

Let's examine these select problematic areas of safety compliance...

"Safety focused" organizations have maxed out compliance already

Where is your organization currently out of compliance with something safety related?

You either know that you are compliant because you are frequently looking at compliance related things, or you know that you are out of compliance (and working rapidly to remedy the issue) because you are frequently looking at compliance related things. Either way it tells us the same story – you are really great at compliance. You are so great at it that you have maxed out the return on your compliance efforts – doing more of it, or doing compliance harder or better, will not yield a better result than you are already experiencing.

Catastrophic events still occur in compliant organizations

The creation, maintenance, and enforcement of safety compliance simply does not prevent significant or catastrophic outcomes. In fact, if blind rule following or compliance persists in the face of cues that suggest that adaptations should be made, this may lead to undesirable or catastrophic outcomes (Dekker, 2014).

Current OSHA fatality information highlights this occurrence of catastrophic events in "compliant" organizations. A quick examination of the 80 most recent fatal events listed (at the time of publishing) on The Occupational Safety and Health Administration's website reveals that 48 out of the 80 employers of those killed in these events were cited for some variety of safety non-compliance after the event (OSHA, n.d.-c). So, let's say that roughly half of these companies that incurred a fatal event were "in compliance" at the time of the event – this being evidenced by the lack of citation. How likely is it that non-compliance truly played a part in the other 50%? Who knows… Either way, it is a clear indication that even "compliant" organizations – at least compliant enough to avoid an OSHA citation after a fatal event – still kill people.

Safety becomes a bureaucratic activity to be demonstrated "up"

Our overfocus on compliance and on safety management systems in general (systems that typically contain a large element of compliance focus at their core), has given rise to bloated safety bureaucracies and cultures of compliance (Dekker, 2019). These 'cultures of compliance' drive employee behaviors towards the completing of compliance "things," rather than the investing of time into things that truly matter to the safety of work. Rather than spending time focused on things vital to safety, employees find themselves tasked with things like pre-audit walkdowns, compliance audit readiness meetings, and the excavation of years-old safety paperwork to be able to demonstrate compliance. When asked, "is your location in compliance" from above, the only acceptable answer is "Yes, always!"

A story about blame transference…

In some recent travels, I was spending some time with an organization that was a few years into their Human and Organizational Performance journey. They had done a really great job with it over the first few years and had made quite the positive impact on their company. One item they were particularly proud of (and rightfully so) was their shift away from focusing on OSHA recordable events, and their massive shift away from the application of blame after an accident. Anyone that has been a part of – or sought after – the transformational change that Human and Organizational Performance brings about, knows that both of these are watershed moments in your journey and incredibly hard (initially) to bring about. They were proud, and I was proud of them.

As we set in a large conference room catching up and talking about their HOP journey, I could not help but notice a long list of bullet points scribbled on the whiteboard across the room. I am naturally curious, so I could not help myself – mid-sentence I made my way across the room to take a quick peek. It was a long list of "common safety audit items." It included things like caps on eyewash stations, missing knockout plugs, faded signs, missing signatures on paperwork, and on, and on… I had to ask, "what's all this?"

Now, you might assume at this point that I am quite anti-compliance, which would be far from the truth (more on that in a moment). I was honestly expecting to hear a story about how these were items that had been brought up as areas for improvement, a focus list of things that commonly degrade and need replaced within their particular operations, or some notes that had been left over from a recent learning team, but what I heard was quite different. I listened to these leaders describe the painful situation they currently found themselves facing – a story about the overvaluing of safety compliance, and a story about the application of blame.

These mid-level managers described how leaders up through their company would panic, breakdown, meltdown, and throw executive temper tantrums over findings related to safety audits or assessments. And just like that, with a little scratching beneath the surface, we discovered that our traditional approaches to safety had once again reared their ugly head. Now, having outgrown their desires to panic, breakdown, meltdown, and throw executive temper tantrums over events, they had placed this reaction elsewhere – poor assessment and audit findings were now acting as a surrogate for events and accidents. Blame had not been eliminated from this organization; it had simply been moved.

Be cautious to not allow blame to transfer to other "safety things" while you are in transition away from blaming around events. Remember, blame fixes nothing (Conklin, 2019). Allow me to expand upon that by stating that blame fixes nothing – full stop. It does not fix events, problems, or even compliance audit findings – it only moves us farther away from learning and fixing.

Thinking about compliance differently...

How should we be looking at, and thinking about safety compliance? It is certainly not through "compliance or else" approaches, or through an overfocus and overvaluation on the act of compliance itself. I do believe that compliance is a good thing – it has helped us come a very long way over the years – I just do not believe that doing it harder or with more rigor will give us a better result. We have maxed out the return on our compliance investments. That is not to say that we should go around ripping required guarding from machinery or pulling handrails from stairs – we definitely should seek to maintain the better work environments compliance has helped us create – but we should begin to think about compliance a bit differently. As one of these leaders from the previously mentioned conversation stated, "It'd be nice if they could just see these findings for what they are, an opportunity to fix something that is broken."

What a difference in approach (and in reaction) that would make – seeing safety audit findings as a gift, rather than a curse. This leader was spot on in their thoughts, we must move safety compliance (and the auditing of it) along with us on this Human and Organizational Performance journey. The emergence of safety compliance things in our work worlds – similar to the manifesting of undesirable behaviors – is not the problem, it is a symptom of a problem. The

237

overvaluing of these symptomatic issues without deeper exploration into where they grow from will not render our workplaces better – it might actually make them worse.

We must let go of this notion that through more and more compliance we will create safety – a notion that leads us to react poorly upon the discovery of non-compliance due to the assumption that "non-compliance" means "unsafe." Rather than our typical reaction of "oh shit!" when we discover the existence of compliance issues, we must shift that reaction to "good." We must begin to see safety compliance, not as a measure of how "safe" our companies are, but as the bare minimum for any habitable workplace – as a starting point. We must begin to see safety compliance findings, not as measures of how "unsafe" our organizations are, but as opportunities to learn more, to fix things, to get better, and to grow.

Stop trying to comply your way into safety excellence – that has never worked, and it never will. Compliance will never guarantee safety, but an overfocus on compliance will certainly lead you away from focusing on the things that help create better and safer outcomes within our work worlds.

KEY POINTS

"Safety focused" organizations have maxed out compliance – complying harder does not create safer organizations

Catastrophic events still occur in compliant organizations

An overfocus on compliance results in safety becoming a bureaucratic activity to be demonstrated "up"

View audit findings as a gift, rather than a curse

Stop trying to comply your way into operational excellence – that has never worked, and it never will.

PUTTING IT INTO PRACTICE

- Focus on tempering reactions to 'poor' audit findings – embrace responding with "good" rather than "oh shit!"
- Retune auditing and assessing to examine more for effectiveness than compliance
- Focus auditing and assessing efforts on the presence and viability of lifesaving controls and safeguards

10 IDEAS TO MAKE SAFETY SUCK LESS

REDEFINE

"SAFE"

CONVERSATION PRIMERS

How does your organization currently define "safe?"

How much time does your company invest into the prevention or minimization of low outcome events?

If a significant amount, why?

10 IDEAS TO MAKE SAFETY SUCK LESS

What does "safe" mean to your organization?

I had a close friend reach out to me recently seeking advice regarding some safety issues he was facing at work. This particular friend works as a mid-level leader for a massive construction and maintenance contractor and has been doing this type of work for most of his adult life – starting in the trades and slowly working his way into leadership over the years. I have known this person for quite some time – having worked together on several projects in the past – and knew that this must surely be a serious issue for him to be calling about something work-related out of the blue. We met for coffee a short while after our initial phone conversation to dig into the meat of the problems he was facing, and to see just how I could help.

After a brief exchange of pleasantries – time spent catching up on any personal "life" things that had occurred since the last time we had seen each other – he dove right into the matter at hand. "I have serious safety problems at work," he inserted abruptly to steer our personal conversation back towards the work dilemma he was currently facing. "We had 12 events last year, and 7 so far this year," he stated. "Corporate is freaking out and they are demanding that I do something to make sure that we do not have any more of these," he continued. He then went on to describe what seemed like an endless and extremely detailed list of "corrective actions." I stopped him and asked, "Hold on, let's back things up a bit. Can you tell me more about these events?" Now, certain that I would hear several stories about life altering events, I readied my pen and paper to take notes. "That's the problem…" he said with obvious disgust. "They're first aids – simple bumps, scrapes, and bee stings – I have no clue how to prevent those," he continued frustratedly.

He then went on to describe a work world that is all too familiar for many working in "safety focused" organizations and industries. He described a company infatuated with prevention – one with an overt focus on the prevention of bumps and bruises thinking that, through these preventative efforts, they were preventing more catastrophic events in the future. He highlighted their mission of "getting to ZERO," how leaders up through their organization would react poorly to events, how they would painstakingly pick apart and "armchair quarterback" even the most minor of first aid events, and how all of this focus on the trivial was forcing his team to attempt to case manage first aid injuries (to manage them out of being "first aids" as defined in the OSHA standard), or to simply stop reporting events up through the organization altogether.

He described how the company executives had labeled the previous year as "the worst year ever for safety" due to the 12 first aids, how he had been forced to withhold employee bonuses, how he was working overtime trying to pull together hundreds of pages of event paperwork, and how he was now being forced to administer discipline (as part of the "fix") after events. My dear friend was neck deep in a work world comprised of safety madness.

We spent the next several hours constructing a better "safety plan," one that had nothing to do with that handful of bumps and scrapes but had everything to do with doing safety a bit differently.

So, what ultimately happened? His plan was scrapped. Despite his loud and constant pleas for change, and his begging to move in a better direction, the company was just not ready to move in a better direction – they were not ready to let go of their current definition of safety.

Traditional definitions of "safe"

Traditional definitions of safety – or descriptions of what is "safe" – describe an environment that is free from unwanted safety events. "Safe" is typically viewed as a state of being or as an end-state devoid of accidents – one that is evidenced by a lack of negative occurrences – that is commonly referred to as "zero harm," or simply "zero."

In these "zero harm" approaches…

-we assume that-

All incidents are preventable

And that by…

Closely examining and preventing small events we can predict and prevent big events on the horizon

With these assumptions in mind, organizations work feverishly to render their workplaces "safe" by bringing their total sum of safety accidents to zero. Many "zero harm" focused companies do indeed reach this target – you do not have to look very far to find organizations touting millions of manhours worked without a scratch or bruise, or companies celebrating hundreds upon hundreds of days since their last OSHA recordable. These organizations point to this lack of negative events as an indication that their work worlds are "safe," but are they?

Research into the efficacy of these 'zero' approaches to safety reveal that they are wildly ineffective in the prevention of larger and more catastrophic events.

A 2017 study of the top 20 U.K. construction companies revealed that companies with an explicit 'zero' policy actually suffered more major events and fatalities than those without a 'zero' policy – demonstrating the potential existence of a 'zero paradox' (Sherratt & Dainty, 2017).

In a 2000 study within aviation, Barnett and Wang demonstrated that passenger mortality risk is the highest in airlines that report the fewest number of events (Barnett and Wang, 2000).

In commentary relating to the literature around 'zero,' Sidney Dekker highlights that the research points towards the fact that 'zero' does not prevent fatalities or major accidents. In fact, parts of the literature demonstrates that a reduction in minor events increases the risk of major accidents and fatalities (Dekker, 2017).

Additionally, recent research demonstrates that total recordable incident rate (TRIR) is not a good predictor of more serious injuries or fatal events. This research concludes that the occurrence of recordable injuries is almost entirely random and not predictive of more catastrophic future events (Hallowell et. al, 2020).

In addition to the literature, we see 'zero harms' lack of effectiveness occurring in many real-world, recognizable, and tragic examples such as:

Deep Water Horizon

The explosion of this offshore oil drilling rig in 2010, killed 11 workers and injured dozens more along with releasing 200 million gallons of oil into the Gulf of Mexico. The rig owner, Transocean, had a "strong overall" safety record with no major incidents for 7 years – Deepwater Horizon had even received an award for its safety performance in 2009 (ABC News, 2010).

The BP Texas City Explosion

The BP Texas City Refinery explosion occurred in 2005 when a vapor cloud of natural gas and petroleum violently exploded killing 15 workers, injuring 180 others, and severely damaging the refinery. Investigatory findings revealed total recordable incident rates (TRIR) and lost time incident rates (LTIR) do not effectively predict a facility's risk for a catastrophic event (CSB, 2007).

The evidence of 'zero' being ineffective on its best days, and harmful on its worst, seems quite clear in the current (and ever-growing) body of research around this topic of 'zero harm.' 'Zero,' while being a noble goal, seems like a horrible target – one that leads organizations towards a fixation on the minor, and away from a focus on the meaningful. The application of these 'zero harm' approaches to safety appears to create organizational silence around events and hazards, moving companies farther away from vital operational intelligence and learnings. 'Zero harm' – while it sounds great on paper – is a horrible idea in practice.

To balance out this exploration into 'zero harm,' there are some clear "pros" – as noted by Dekker in *The Field Guide to Understanding 'Human Error'* – relating to the implementation of a 'zero harm' approach to safety:

- Leads to significant cost-savings on healthcare, insurance, and other compensation costs

- Creates a better chance of getting contracts renewed or securing additional work

- Reduces the likelihood of regulatory inspections (Dekker, 2014)

But at what cost do these benefits come? They arrive through the creation of things like safety secrecy and bloody pockets – cost savings that are ultimately born out of people keeping their mouths shut and carrying injuries home with them. These are hardly real benefits when weighed against the problematic mechanisms that bring them about – organizational behaviors that commonly emerge in response to a targeting of 'zero.'

Now, 'zero' champions and zealots will be quick to callout that these are simply occurrences of "zero gone wrong," "zero not used correctly," and that if, "zero had only been properly applied" in these particular situations, then the result would have been vastly different. For the purposes of this book, we will remain in the reality of how 'zero' is commonly applied within organizations – we will not venture down the presumptive path of "zero applied better."

Typical organizational approaches to 'zero'

Within our work worlds, what does it mean to apply 'zero harm' approaches to safety? Keeping in mind that these approaches' view "safe" as a lack of events, it is easy to trace out what these typical zero-based safety management systems hope to do – 'zero' hopes to manage safety outcomes.

'Zero harm' is a goal, it is a target, to create an incident and injury free workplace. The hopes of 'zero' is to bring all safety related events to an end, believing that by doing so, the workplace is rendered "safe" and free from the likelihood of higher outcome events. Some of the most common organizational approaches to 'zero harm' are clearly demonstrated in the internal company languages these 'zero' focused organizations speak – things like "if you focus on the little things, the big things take care of themselves," as an example.

This belief that managing of the "little things" prevents larger and more catastrophic things in the future, leads organizations towards a focus on the "little things" they believe to be causal of more significant events. 'Zero' drives organizations to meticulously count bumps and scrapes, invest ungodly amounts of time into investigating first aid events, and yes, it even drives them to call a year, "the worst year ever," because of a handful of back sprains and bruises.

This current definition of "safe" – the lack of accidents in the workplace – drives a primary focus on numbers and targets, it causes organizations to obsess over the minor events that make up these numbers, and it biases them towards a sole focus on prevention.

This extreme focus on numbers and metrics results in organizations taking just as extreme actions to manage (or manipulate) these targets and metrics. We will often find the tying of large safety bonuses to these metrics within 'zero' organizations – basically bribing leaders and the workforce to remain silent around anything that might reduce the likelihood of receiving a bonus.

This type of manipulation was painfully demonstrated in a 2012 major fraud case in the United States. A safety manager for a large maintenance and construction contractor was sentenced to 78 months in prison for providing false information about injuries through the underreporting of their numbers and severity. The evidence presented at trial encompassed over 80 injuries, including broken bones, torn ligaments, hernias, lacerations, and shoulder, back, and knee injuries that were not properly recorded. Why? To collect safety bonuses worth over $2.5 million from a client (Department of Justice, 2013).

Beyond a fixation with numbers and metrics, these 'zero harm' definitions of "safe" promote shallow or surface-level fixes of safety problems and promote the blaming of workers for unwanted events.

How this usually plays out at work…

The little things like bumps and scrapes lead to more significant events like amputations and fatalities. If we have (insert some arbitrary number here of) hand lacerations, then at some point we'll have a fatal event because of a hand laceration. So, we must predict and prevent all of these to prevent a fatal event.

We'll usually take this further to say….

We have too many hand lacerations, if we do not get these numbers down, then we'll eventually have a hand laceration fatality! Poor behaviors on the part of our workers are at the heart of this issue… We must do a stand-down, mandatory hand safety training, acquire bright-pink gloves so employees are more aware of their hands, we must introduce and incentivize a hand safety metric, we must draft additional hand safety rules, and we must severely punish any wrongdoer that we find!

'Zero," through its primary assumption that all incidents are preventable, moves us quickly towards blaming workers for any unwanted events. This is especially true relating to

257

extremely minor safety events such as bumps and bruises – events that are easily argued to be unpreventable – due to the lack of any other obvious preventative approach that could have been applied to the event. Without a clear and obvious prevention method that could have been used to eliminate the occurrence – while firmly clinging to the belief that all incidents are preventable – organizations quickly shift towards placing blame on the worker.

These high probability low outcome events also drive organizations to do some pretty wonky things in the name of "preventing all events – even bumps and scrapes" as it relates to "corrective actions." After an extremely minor safety event – still believing that everything must be prevented – organizations will embark on "safety prevention plans" chocked full of shallow actions such as retraining, standdowns, more rules, more observations, and harsher punishments.

Our current definition of "safe" that describes an environment that is free from all unwanted safety events, one that is evidenced by a lack of negative occurrences and is commonly sought after through the application of 'zero harm' approaches to safety, is simply not working.

The redefining of "safe"

Rather than viewing safety as the absence of events, safety is better defined as the presence of defenses (Conklin, 2012). This wildly different definition of what "safe" means, takes us in an entirely new direction than our previous description.

Defining "safe" as the presence of defenses propels us towards a focus on what is meaningful, it moves us away from viewing safety as an outcome (a number) to manage, it forces us to let go of this misguided and harmful notion of 'zero,' and it leads us to form better assumptions relating to the safety of work.

We simply cannot manage safety outcomes – seeking to manage downstream results never changes much anyways – but we can manage the presence of defenses that allow for better outcomes. With some better assumptions that tell us things like failure will occur, people will make mistakes, safety is the presence of controls, and on, we can seek to influence what matters – we can seek to create an environment in which people can fail as safely as possible. This better definition of "safe" leads us away from our prevention bias and pulls us towards an organizational mindset of "assuring work goes well," rather than attempting to "make sure that work does not go wrong."

While these shifts may seem small, do not let their perceived simplicity fool you. These are powerful and paradigm shifting ideas. The difference between "can't fail," "might fail," and "will fail" is massive – it is a massive shift in organizational beliefs. Our bias towards prevention has left us in the "might fail" mindset for far too long. Our focus on more traditional approaches to safety have left our organizations truly believing that, with enough safety focus and effort, then things just can't go wrong. "Can't fail" and "might fail" are dangerous waters to find yourself in. With the better assumption of "will fail," we naturally gravitate towards seeking out ways to fail a bit softer, ways to diminish the outcomes of events from catastrophes to bruises, we move towards the idea of getting good at failing gracefully, and we move in the direction of strong defenses and lifesaving controls.

Viewing safety as the presence of defenses, rather than the absence of accidents, is a vital watershed moment in your Human and Organizational Performance journey – it is crucial to bringing about the change you hope to see. So many other positive changes cascade from this shift in beliefs. But you must be willing to outgrow a bias towards prevention – you must be willing to part ways with 'zero' – and you must be willing to fully embrace this new definition of safety that tells us…

Safety is the presence of defenses.

KEY POINTS

The application of 'zero' based safety strategies often causes more harm than good

'zero' based approaches to safety are wildly ineffective in the prevention of larger and more catastrophic events

The current definition of "safe" drives a primary focus on numbers and targets

Rather than viewing safety as the absence of events, safety is better defined as the presence of defenses

Defining "safe" as the presence of defenses propels us towards a focus on what is meaningful, and it moves us away from viewing safety as an outcome to manage

PUTTING IT INTO
PRACTICE

- Transition away from 'zero' based safety
- Redefine safety within your organization
- Lessen organizational focus on numbers, rates, and metrics, along with removing safety incentives and targets
- Refocus organizational efforts on examining for the presence of defenses

10 IDEAS TO MAKE SAFETY SUCK LESS

GIVE UP
ON SAFETY
"FORTUNETELLING"

CONVERSATION PRIMERS

How much focus does your company place on the predicting of events?

If a significant amount, what types of data do you use for these efforts? How accurate are you?

What part does prevention play in your current approaches to the safety of work?

10 IDEAS TO MAKE SAFETY SUCK LESS

Safety fortunetelling - *the act or practice of predicting the future relating to workplace safety.*

Safety fortunetelling is commonplace within many high-risk organizations. It is used as an attempt to peer into the future state of an organization to predict what events or injuries are going to take place, when they are going to occur, and where they are going to happen. Organizations take up this fortunetelling to support their efforts around overall accident prevention.

10 IDEAS
TO MAKE
SAFETY
SUCK LESS

We have to start this conversation with a question. What are we trying to predict with all these safety fortunetelling efforts? Now, many will give the vague run-of-the-mill answer of "all safety events" – a response that wreaks of 'zero'

– but "all safety events" is what is truly meant. Most safety fortunetelling efforts are built around this idea of predicting – based off things like safety observations, assessments, audits, behaviors, and past incident data – so that one might prevent unwanted or undesired outcomes altogether and yield the desired result of 'zero.'

In more traditional approaches to safety, these unwanted or undesired outcomes can be described as any safety related event – close calls, first aids, recordables, lost time events, and up to and including fatalities and catastrophes – anything that is more than 'zero.' In these more traditional approaches, lower outcome events are often viewed as being predictive of larger and more catastrophic things yet to come. This props up a belief of "managing the little things…" – of removing the future potential for catastrophic events by predicting and preventing lower outcome occurrences.

When significant safety events do occur, we are quick to look back on these events (and the lower-level occurrences we believe to be predictive of them) and ask, how did we fail to predict this? How did we fail to prevent this? Believing that all events are predictable and preventable, and that the goal of any "good" safety program should be primarily built around this idea of predicting and preventing unwanted outcomes, pushes organizations to favor this notion of prediction and prevention above all else. Believing firmly in the predictive capacity of things like lower-level events (commonly referred to as lagging indicators) and things like behavioral observation data, compliance audit findings, pre-job brief use, and similar – organizational data points usually coined "leading indicators" – leads many organizations to see the greater collection and analyzing of this data, or the placing of more effort and rigor into prediction and prevention as a corrective action after the occurrence of a significant event.

These ideas turn organizations into massive data collecting machines, hoping to capture every minute tidbit of information in hopes that it is (or could be) a vital piece of this predictive puzzle. We love safety data, these numerical representations of "safety" in our work worlds, these clues that we believe will help us solve this mystery of prediction – they drive us to want more and more. Through the bulk collection of safety data, and by meticulously analyzing and trending this safety data, organizations believe that their workplaces can be rendered "safe" – free from unwanted safety outcomes – by predicting where bad things are about to (or likely to) occur and swiftly act to prevent them.

Our work worlds are biased towards prevention, so much so that we rarely think beyond it. Prevention is a great thing and tons of great innovation has happened in that space – things that have resulted in a safer world all around – but how much farther can prevention alone carry us? Worse yet, at what point does this desire to "predict and prevent" everything become harmful?

Some problems with our desires to predict and prevent

Humans are notoriously bad at predicting the future – we are just flat out terrible at it. To compound this issue, we are wired to skew towards what we would like to see happen. Research suggests that the more desirable a future event is, the more likely people think it is (Eiser & Eiser, 1975). Conversely, the more someone dreads or fears a potential outcome, the less likely they think it is to happen. As an example, in November of 2007, economists in the *Philadelphia Federal Reserve's Survey of Professional Forecasters* predicted just a 20 percent chance of "negative growth" in the U.S. economy any time in 2008, despite visible signals of an impending recession (Beaton, 2017). What followed was the most severe economic recession in the United States since the Great Depression of the 1930s (Investopedia, 2022). Relating this to our safety predictions, we often assume that our preventative efforts are robust, that our processes are good, that our organizations are "safe," and ultimately, that they cannot fail – we slant towards optimism and overconfidence while avoiding the fact that all of these preventative efforts will – at some point – degrade, breakdown, and fail.

Another issue with this bias towards prediction and prevention is that it moves us towards "if something bad happens," and away from "when something bad happens." So often, we seem to place greater importance on safety data than safety reality – we lose touch with the reality of our work worlds. We believe that – with enough focus on prevention – we eliminate the potential for catastrophic events in our workplaces. And that, if we can just get enough safety data flowing in, we can see or predict when that next fatality or catastrophe is about to occur – allowing us ample time to bolster our preventative efforts and keep the unwanted outcome from occurring.

To make matters even worse, catastrophic events are especially hard to predict. The problem? Serious injuries and fatalities are outliers, they are not normal, they are anomalies in our systems (Conklin, 2017). An anomaly, being something that deviates from what is standard, normal, or expected, makes them highly unpredictable. As we have seen in previous chapters, our traditional "predict and prevent" approaches to safety simply do not work for more extreme events.

Redefining our goals....

When talking about safety data in particular, the car analogy is likely a bit overused – but I will share it with you anyways. If you thought that you could read a book about safety and avoid hearing a story about driving, you thought wrong...

I tend to think about all of these various data points – these "leading" and "lagging" indicators – similar to how you would interpret feedback while driving a car. In many organizations, we try to operate our "cars" through the rearview mirror – we try to see where we are going by closely examining where we have been. After we hit a wall or run off a cliff one too many times, we typically adjust our approach. Rather than trying to drive through the rearview mirror, we now become fixated on the dashboard. We try to drive our car by looking at the various gauges, measures, and bits of information relating to the operational state of our car. After we hit a road sign or run into a ditch, we will again adjust our approach – we'll start trying to drive our car by not only staring at the dashboard, but we will now glance in the rearview mirror occasionally. Yet again, we crash.

Now we are frustrated, and we set out to fix this problem! We have leading and lagging indicators already, so what should we do? We need better leading and lagging indicators – better gauges and measures, and bits of information along with better analysis of what has already passed us by. Surely, with all of these extra indicators and a heightened focus on what we have already driven past, we will get better at driving our car. This time we hit and kill a pedestrian trying to cross the road.

The point of this story is not to belittle the importance of our gauges and dashboards – it's quite a good thing to know how much fuel you have, or if your car is overheating. I'm not saying that we should not look back and learn from the events we have experienced – that is definitely a great thing to do. But trying to drive our cars while staring in the rearview mirror, at the dashboard, or both, isn't a really great way to drive a car. What is the most important part of a moving car? The stuff that is in front of the car. Is it ok to glance in the mirror? Sure, that is probably a good idea. Should we keep an eye on those gauges? Absolutely. But, most important of all, we need to be looking through the windshield – we must be looking at reality.

So, we find ourselves at another question concerning our fixation on predicting and preventing safety events – what is it that we are hoping to achieve? If our goal is to stop maiming and killing workers, predicting and preventing is not the path forward – we have gotten about as good at that as we're going to get. Trying to predict and prevent "harder" likely causes organizations to kill more workers, not less.

It's not that traditional safety is garbage – our traditional approaches are pretty effective for a lot of things. Good work is good work, and effective is effective. If it is good and effective work, and it aligns with our Human and Organizational Performance principles, we should probably keep those efforts up. But we have to stop fooling ourselves (and our organizations) into believing that we will eventually predict and prevent our way out of failure. Failure is the constant – the one thing we can depend on…

Let's take a moment to highlight a few key ideas…

Some key ideas:

- We are not good at predicting events – especially fatalities
- Prevention is a good thing, but a sole or overfocus on prevention is harmful
- Increased focus and effort on predicting and preventing will not yield a better result

While we should not give up our efforts around prevention – it is really effective for a lot of things – we have to begin to understand that prevention alone is not effective at reducing serious injuries and fatalities. We must do things differently if we want to see a different result. Continuing to double down on predicting better and preventing more, will only doom us to an existence of "more of the same."

Serious injuries and fatalities are unpredictable events that exist within uncertainty, and we can't predict fatalities or completely control uncertainty (Conklin, 2017). We cannot manage uncertainty, but we can manage the presence of controls. We cannot predict and prevent every potential bad thing that could occur, but we can create a resilient system that can effectively manage failures.

Rather than wasting our time on safety fortunetelling – seeking to predict and prevent everything – our time is far better invested in designing systems that will not result in catastrophic outcomes when they fail.

KEY POINTS

Safety fortunetelling efforts are undertaken in an attempt to predict and prevent events

> *We are not good at predicting events – we simply cannot see into the future*

Prevention is a good thing, but a sole or overfocus on prevention is harmful

> *Increased focus and effort on predicting and preventing will not yield a better result*

Rather than wasting our time on safety fortunetelling, our time is far better invested in designing systems that will not result in catastrophic outcomes when they fail

PUTTING IT INTO
PRACTICE

- Eliminate safety fortunetelling efforts from your organization
- Do not give up prevention, but seek to grow your organizations focus beyond it
- Invest time and energy into designing systems that will not result in catastrophic outcomes when they fail
- Focus organizational efforts towards management of safeguards and lifesaving controls

10 IDEAS TO MAKE SAFETY SUCK LESS

EMBRACE HUMANITY

CONVERSATION PRIMERS

How does your organization typically view workers?

What does the term 'human error' mean to your company?

Do you view error as a choice?

10 IDEAS TO MAKE SAFETY SUCK LESS

We dehumanize workers. Does that sting a bit? It might, but it's true. That is so much of what we have talked about throughout this book, how organizations can move beyond typical dehumanizing management approaches and tactics, and fully embrace the humanity of those in their care.

Treating employees like unruly children, being untrusting of them, inundating them with endless rules, forcefully doing things to them, constantly surveilling them, and much more – all of these things work against human qualities, are demeaning, and rob workers of their dignity. This dehumanization of workers only creates unhappy, disengaged, and ineffective employees, while minimizing organizational learnings, killing company innovation, and crippling safety and efficiency.

Organizations dehumanize workers because they view them as a problem to control. People are viewed as problem creators – causes of trouble within organizational systems – rather than problem solvers. An assumption is made that organizational systems would be fine if it were not for the unpredictable and erratic behaviors of unreliable workers (Dekker, 2014).

Organizations embrace and lean into this notion –
viewing people as the problem – and point to
human missteps or 'errors' as causal of things like
events, injuries, and other operational surprises.
This element of 'human error' – since
organizations always find it after an event or
operational surprise – becomes fodder for efforts
seeking to 'fix' workers, eliminate errors, and
ultimately, cure the workforce of their
bothersome humanity. 'Human error' becomes
the boogieman, and the elimination of this
boogieman becomes the futile quest of the
organization.

Some problems with clinging to 'Human Error'

Let's run some of this 'human error' logic to
ground – when we point to 'human error' as the
culprit, what are we really saying about those that
have erred? Beyond the stating of an obvious fact
that error occurred, 'human error' is typically a
loaded term. 'Human error' is a judgment against
those that have erred – a judgment heavily
influenced by hindsight bias – and therefore
counter-productive to the understanding of things
that have gone wrong (Besnard & Hollnagel,
2012) (Woods et. al 1994).

When 'human error' is viewed as causal, what is
really being said is that something bad happened
because someone – typically a worker near the

coalface – messed things up. 'Human error' is framed as a choice – to err or not to err – made by the worker. Since error is viewed as a choice, and because 'good workers' would never choose to err, 'human error' becomes synonymous with blame. What is really being stated is that 'good workers' do not make mistakes, 'bad workers' make mistakes – bad things happen to bad people. To get to the core of this rotten logic, a typical belief is that, if workers were better people, bad things would not happen to them.

We clearly demonstrate this belief in our responses to 'human error.' We blame, shame, and retrain, hoping to turn our employees into better humans. We punish, beat, and banish workers trying to convince them to stop choosing 'human error' over success. Organizations view 'human error' as being a poor choice on part of the worker – a 'choice' with harsh and painful consequences – and demand that workers make better choices as to avoid unwanted outcomes.

Letting go of 'human error'

As defined by Conklin, error is the unintentional deviation from an expected outcome (Conklin, 2019). How can an unintentional deviation be a choice? It can't. The thing about error – and the events occurring after an error – is that they are unexpected outcomes. Everything makes perfect sense, everything seems to be going according to plan, work seems to be going well, until suddenly it's not. If employees can foresee that things are going to go badly, they will not proceed. 'Human error' only appears to be a choice with the gift of hindsight and known outcome (Dekker, 2014).

"To err is human…" as the old saying goes – it's spot on by the way. But so is to adapt, figure out, work through, adjust, and all of the other key elements that we rely upon to generate successful outcomes within our work worlds and beyond. The potential for 'error' is baked into our D.N.A., along with all of the other things that make up our humanity – that give us the "human element." And guess what, people usually get it right. People are typically successful at the things they set out to do.

Instead of viewing 'human error' as causal of unwanted outcomes, we should seek to understand why the same behaviors – behaviors only viewed as 'errors' after an unwanted outcome – typically make things go right and occasionally make things go wrong (Besnard & Hollnagel, 2012).

Embracing the human element

If our desire is to learn deep and contextual information about normal work, we must let go of this notion that 'human error' is the cause of events within our work worlds. Citing 'human error' only holds us back, preventing us from digging deeper into how unintended operational surprises actually manifest in our organizations.

That is so much of what these 10 ideas focus on – the clearing of roadblocks to learning. It is moving beyond the things that stand in the way of learning, so that we can learn more about areas of critical risk, so that we can create systems that will not result in catastrophic outcomes when they fail, and so that lifesaving controls and safeguards can be maintained and improved upon.

The only choice in human error is the choosing of how we view it. Do not allow misguided assumptions around 'human error' to stand in the way of deep and meaningful operational learning that leads to the bettering of lifesaving controls and safeguards.

KEY POINTS

Traditional safety management views of workers dehumanize them

'Human error' becomes the boogieman, and the elimination of this boogieman becomes the futile quest of the organization.

Error is the unintentional deviation from an expected outcome

Error is not a choice

If our desire is to learn deep and contextual information, we must let go of this notion that 'human error' is the cause of events

Deep and purposeful learning leads to overall system betterment, and will help maintain and improve upon lifesaving controls and safeguards

PUTTING IT INTO
PRACTICE

- Focus on shifting organizational assumptions about 'human error'
- Remove the option of 'human error' as a cause from organizational processes
- Modify organizational tactics around 'event investigations' – shift them towards learning reviews
- Seek to better system/setting rather than attempting to modify human behavior through reward and punishment

10 IDEAS TO MAKE SAFETY SUCK LESS

WRAPPING THINGS UP WITH SOME FINAL WORDS

So, you have decided to do things 'differently' in your organization – you have decided to bring these ideas to life within your particular workplace – what now? While I hope to have provided you with some practical thoughts on how to accomplish that – ideas sprinkled throughout the preceding chapters of this book – I want to leave you with some ideas relating to an overall organizational shift towards Human and Organizational Performance.

While deep and fundamental organizational change efforts can be quite the daunting task, do not be fearful of taking up the challenge of making things better. Yes, these efforts take time and sometimes move painfully slow. Yes, of course, the organization will sometimes regress or fall back to their 'old ways.' Yes, you will encounter leaders that are just not onboard or actively seek to derail these efforts. But these are all just points along the way – steppingstones of organizational betterment – leading you towards a better workplace. Remember, this is a journey...

As with any journey, you will hit bumps, jumps, and roadblocks along the way. Embrace the process – accept the process. I myself struggled with some of these challenges in my first experiences with bringing HOP to life – I particularly struggled with the "moving backwards" from time-to-time piece. When you find yourself feeling frustrated, simply 'zoom out' and look at the bigger picture. Take a step back and take in where the organization has been, where it is at, and where it is going – the amount of positive change you will see, will often surprise you and ease your frustrations.

Because your organization is unique, your journey will also be extremely unique – it should be. Take these ideas, these concepts, and thoughts on bringing Human and Organizational Performance to life and create a bespoke approach to effectively bring about positive change within your work world.

Do not approach these concepts with a traditional mindset

So very often I see companies attempting to 'force fit' Human and Organizational Performance into their organizations, attempting to meticulously plan every step of this journey onto a timeline of 'HOP implementation,' or trying to 'do HOP' using the same organizational methods and tactics they have used for much of everything else – using an approach akin to the rolling out of a 'safety program.' But Human and Organizational Performance is not a program and approaching it like one only creates headaches and problems along the way – it only stifles progress or leaves you with some bastardized end product far from the true intent of these concepts and ideas.

These concepts and ideas are different, so we must approach them differently. Be very cautious of typical organizational desires to simplify, standardize, and force fit to create progress and change – these methods always backfire. Human and Organizational Performance is a set of beliefs that shape our programs, tools, behaviors, and language (Baker, 2019) – it is not a program to 'roll out.' We simply cannot shift beliefs through the application of a program, we cannot just "roll out" new assumptions into our organizations, we cannot bring about this change by trying to force fit it in. You must grow Human and Organizational Performance from within your organization by reshaping organizational assumptions and beliefs around error, blame, learning, the definition of safety, and on...

Human and Organizational Performance is not a program, but you should have a plan. You need a blueprint; you need a recipe for the cake. You need to bring the right ingredients together at the right times – you don't want to be going for a moist and delicious chocolate cake and end up with a tart – no one wants a tart. You need to have the right people working on the right things at the right times. You need to figure out what your little HOP army looks like, how that works, where your 'bright spots' are, where you are going, and how you think you're going to get there. Do not get completely tied down by planning – this plan should never be rigid. It will move, shift and shuffle – just as it should. Things will get pushed out and other things will get pulled in – but you need a road map to get you in the general vicinity.

A few planning considerations prior to beginning the journey:

Organizational Readiness

Where is your organization currently at? Seek to understand the current state of your particular organization and define where it is that you want to go. This assessment of organizational readiness will allow you to craft a customized approach based off the current reality of your organization. It will help you begin your journey at the right time and allow it to start on a sure footing.

Core Team Creation

Who are your internal champions – those knowledgeable and passionate folks – that will help bring this change about? Find them, get them together, and set them up for success by providing them adequate time, resources, and support to take on this task.

Employee Involvement

How are you going to place your workforce at the center of this change – how are you going to ensure that they have a voice? Be very cautious to not "do HOP" to your organization. You can help avoid this by involving your workforce in these change efforts. Involve your employees, listen to them, learn from them, and be sure that their voices – their ideas and their thoughts – shine through and are shown in the results of these efforts.

It is more of a 'framework' than a plan...

In a 2019 article on the *Safety Differently* website, Andrea Baker describes "5 Phases" of Human and Organizational Performance integration:

- Leadership Interest
- Building HOP fluency
- Operational Learning
- Alignment
- Safeguard Management

Let's explore each of these in a little more depth...

Leadership Interest

Seek to gain leadership support within your organization and find leadership champions or sponsors to leverage while on this journey. These Human and Organizational Performance allies are crucial to the overall growth and success of these concepts within your company.

What this looks like…

- Building relationships with leaders
- Mentoring leaders – especially through challenges or events
- Teaching of HOP concepts to leadership
- Making a case for change
- Possibly bringing in outside speakers to help shift views

Building HOP fluency

This is the education component of your journey – the embedding of these concepts and ideas within your organization. Through the teaching of things like Human and Organizational Performance fundamentals, learning teams, and more, you will establish a base-level of knowledge around this new approach. Over time you will begin to notice subtle changes in the language of your organization – your organization will begin to sound like a HOP focused company – your company will begin to "speak HOP."

What this looks like…

- Providing HOP information sessions
- Conducting HOP fundamental training
- Teaching the use of learning teams and learning explorations
- Shifting organizational messaging towards Human and Organizational Performance

Operational Learning

At this point in your journey, you are beginning to embrace tools like learning teams and learning explorations – the organization is shifting towards a deliberate and passionate focus on learning, especially from those that do the work. Do not just seek this learning after an event or operational surprise, go out and 'learn on purpose' about everyday normal work.

What this looks like...

- Starting to use learning teams and learning explorations more and more
- More and more focus on the gaining of context rich information – the old answers (things like "someone messed up") are no longer palatable
- Beginning to see more independent use of learning teams throughout the organization – people will bring you learning teams they did on their own
- Increasing curiosity about context and normal work

Alignment

At a certain point of maturity in your Human and Organizational Performance journey, it will be time to begin to embed HOP principles and operational learning mechanisms into your existing systems, processes, and programs. Sometimes this also calls for a healthy dose of decluttering – the getting rid of things that are counter to these principles, not useful, or no longer needed – to move things forward by the parting of ways with things that cannot be brought into alignment.

What this looks like...

- Altering of processes and programs to bring them into alignment with HOP principles
- The embedding of HOP principles and learning mechanisms into processes and programs
- Decluttering of rules, processes, and programs
- The elimination of rules, processes, and programs that cannot be brought into alignment with HOP principles
- Creating an HOP framework to ensure that HOP is sustainable

Safeguard Management

Now, with these concepts and ideas firmly embedded within the organization, and by using this operational intelligence gained through operational learning mechanisms (such as learning teams or learning explorations) the organization seeks to continuously and collaboratively design, better, and manage safeguards and lifesaving controls.

What this looks like…

- Bettering of existing controls and safeguards
- Bettering of system designs
- Ongoing operational learning around areas of critical risk
- Periodic testing of safeguards and controls

You will find these considerations and '5 Phases' to be crucial when plotting out your organizations journey towards Human and Organizational Performance. Put plenty of thought into these areas as you begin to think about bringing these concepts to life within your workplace, but do not get overly consumed with or tied to rigid planning. There is not "one right way" to bring these fundamental changes about – there is not a true guidebook to making these changes happen. Plot a course and start moving in the right direction, keep your plan flexible, and understand that it will change along the way.

Doing things backwards

I have seen Human and Organizational Performance brought about "backwards" on several occasions – brought to life within organizations with very little leadership interest but using its success to gain leadership interest.

In these cases, Human and Organizational Performance is applied more at a local or group level. These 'bright spots' then act as a catalyst for HOP growth throughout the organization. When the benefits of doing things differently begin to surface, those up through the organization will typically be quick to take notice. This is demonstrating success by doing – doing things differently at a local level, and then pushing those success stories up through the company. Good results are hard to deny, and they quickly lead to more and more excitement and support.

These more localized efforts usually start as somewhat 'grassroots" endeavors – coming to life through the growing of HOP fluency, the letting go of blame, the changing of reactions, the embracing of learning – in a particular subculture of the organization.

While this seems counter to accepted guidance around organizational change efforts – and it is in many ways – I have seen this work well. Especially in organizations that just are not quite ready to take the jump, or with upper-level leadership teams that simply do not see the need for change. These 'backward' approaches can definitely be useful if your organization finds itself wanting change, but without a clear commitment from up within the leadership chain.

Leverage the already mentioned "5 Phases" of Human and Organizational Performance integration (Baker, 2019) while seeking out this more localized approach as well – just apply them in a local fashion. As an example, rather than seeking out executive commitment, this "leadership interest" might look more like support from a local manager, supervisor, or team leader.

An easy place to start…

If it's all just a bit too much for your organization to take on all at once, I often recommend starting out by conducting a few learning teams or learning explorations. Pick an area that could use a little improvement, a particular pain point or problem, or simply choose a job or task that you would like to learn more about and give it a shot. Go out and use these operational learning mechanisms to render your workplace better, and to tell the story of normal work – of reality – up through your organization.

The use of these approaches to gaining operational intelligence are low risk and high reward – they are the perfect opportunity to demonstrate the viability and usefulness of doing things a bit differently.

Temper your expectations

As I have already mentioned, I struggled initially with the overall slowness of change, along with the occurrences of leaders stepping back into more traditional mindsets as I started leading these types of change efforts. It is very easy to find yourself frustrated and disappointed if you do not take time to temper your expectations as you begin on this journey. It is also vital to understand that the indicators of "big progress" within your organization will often be found in the little things.

One of the best places I have found to listen for progress, is by listening to the stories of workers within your organization. When you hear stories of things getting better, of things making more sense, of better experiences – those "little things" are huge indicators of success. Just the fact that people are sharing their stories of 'normal work,' tells you that things are moving in the right direction. When you are feeling worn out and tired, go spend some time listening to the stories contained within your organization.

Stick to the principles

No matter where your organization finds itself on its Human and Organizational Performance journey, always keep the *5 Principles of Human and Organizational Performance* at the heart of your efforts – lean into them, lean into the concepts of *Safety Differently*, and lean into these 10 ideas.

When things get challenging, lean into them that much harder. When things start to move backwards, lean into them even harder. When you find yourself confused or unsure of what to do in a particular situation, allow these principles, concepts, and ideas to guide you – they will not steer you wrong.

Your company is unique…

Your company is very unique, so your journey will also be very unique. Embrace this uniqueness, it's what makes your company great! Your uniqueness should shine through in your plan, and in how you approach bringing these concepts and ideas to life within your work world. Take these ideas and – while sticking to the principles – creatively apply them within your organization. Take these ideas and form fit them to make things work effectively for your company. While leaning heavily into the principles and all of the ideas we have discussed, and while listening and learning from your workforces, shape your organizations very own unique path towards better. Do this and amazing things will follow.

Good luck on your journey! I can't wait to see the awesome things you bring about within your organizations.

EPILOGUE

This journey is worth it.

I have had the distinct honor and pleasure of being involved in various companies' journeys towards the application of these concepts and ideas, and I am here to tell you that zero percent of them have regretted it. I have led these changes while working internally for organizations (that were historically very traditionally based) and had the opportunity to see and feel these changes directly for myself as an employee of an organization growing towards better. This journey is long and slow, but it is worth it.

To see and feel the results of Human and Organizational Performance coming to life within your organization is breathtaking – witnessing the changes quite literally gives you goosebumps. As I reflect back on the first time that I was presented with an opportunity to lead this type of effort for an organization – while I was working for a large organization in the power generation space – what stands out most are the stories.

I heard stories like that of a 30-year veteran of this particular organization describing how, after an event, they were embraced by the company instead of being blamed and fired. I listened to a new employee compare this organization to their last, highlighting the positive difference in their working experience. I heard story after story – too many to share here – each describing how this fundamental shift directly and positively impacted their working lives. These stories are not just 'tall tales,' they are powerful indicators of a move in a better direction. They are a demonstration of bettering the working lives – the lived experiences – of those that reside within our organizations.

What are the stories that currently make up your work world? If you could tap into them right now, what would you hear? Would you hear stories of learning and betterment, or would you hear stories of blame, shame, pain, and employee suffering? Your employees have stories – stories your organization is helping to shape – Are you listening? Are you helping to make them better or worse?

These 10 Ideas will help you bring Human and Organizational Performance to life, they will help you revolutionize your approach to the safety of work (and practically everything else), and they will aid you in bringing about transformational change to your work worlds – through their use you will craft better stories. By bringing these concepts to life, you are deliberately building a better work world and creating a much more positive lived experience for your employees.

The journey is worth it because the people within your organization – those that work tirelessly to get shit done and to get shit done well – they are worth it. Your people, those workers that have been entrusted to your care, they are worth it.

Here's to the constant pursuit of doing things better – to making work (and the safety of work) suck just a bit less – for everyone, especially those getting shit done within our work worlds.

ADDITIONAL RESOURCES

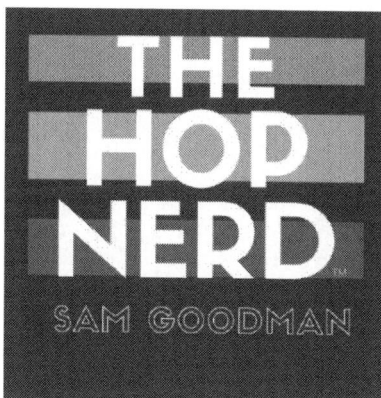

For Human & Organizational Performance training and consulting, learning teams, and safety support visit…

www.thehopnerd.com

Also

Tune into *The HOP Nerd Podcast*, available everywhere you listen to podcasts, for weekly in-depth conversations about doing *safety better*.

A QUICK NOTE

In the following pages you will find a vast number of resources and people to help you on your Human and Organizational Performance journey.

I have attempted to be as accurate as I can – including many amazing thinkers and practitioners – but I am certain that I have missed someone or something.

Any omission has not been on purpose and the lists are in no particular order.

Please take time to learn more about all of these great books, podcasts, and people – you will find these resources invaluable to you along the way.

PEOPLE

Todd Conklin
www.hophub.org

Sidney Dekker
www.sidneydekker.com

Erik Hollnagel
www.erikhollnagel.com

David Woods
www.adaptivecapacitylabs.com

Richard Cook
www.adaptivecapacitylabs.com

Robert Long
www.humandymensions.com

Bob Edwards
www.hopcoach.net

Andrea Baker
www.thehopmentor.com

Dave Provan
www.safetyfutures.com

Teresa Swinton
www.paradigmhp.com

Clive Lloyd
www.gystconsulting.com.au

Jay Allen
www.safetyfm.com

James MacPherson
www.riskfluentltd.com

Mark Alston
www.investigationsdifferently.com.au/

Brent Sutton
www.learningteamscommunity.com

Rosa Antonia Carrillo
www.carrilloconsultants.com

Jeff Lyth
www.safetydifferently.com

Ron Gant
www.scm-safety.com

And many more...

BOOKS

Conklin, T. (2019). The 5 Principles of Human Performance: A contemporary update of the building blocks of Human Performance for the new view of safety. Independently published.

Conklin, T. (2016). Pre-Accident Investigations: Better Questions - An Applied Approach to Operational Learning (1st ed.). CRC Press.

Conklin, T. (2020). When The Worst Accident Happens: A field guide to creating a restorative response to workplace fatalities and catastrophic events. Independently published.

Conklin, T. (2012). Pre-Accident Investigations: An Introduction to Organizational Safety (1st ed.). CRC Press.

Dekker, S., & Conklin, T. (2022). Do Safety Differently. Independently published.

Dekker, S. (2017). The Safety Anarchist: Relying on human expertise and innovation, reducing bureaucracy and compliance (1st ed.). Routledge.

Dekker, S. (2016). Just Culture: Restoring Trust and Accountability in Your Organization, Third Edition (3rd ed.). CRC Press.

Dekker, S. (2014). The Field Guide to Understanding "Human Error" (3rd ed.). CRC Press.

Dekker, S. (2014). Safety Differently: Human Factors for a New Era, Second Edition (2nd ed.). CRC Press.

Dekker, S. (2011). Drift into Failure: From Hunting Broken Components to Understanding Complex Systems (1st ed.). CRC Press.

Lloyd, C. (2021). Next Generation Safety Leadership (1st ed.). CRC Press.

Edwards, B., & Baker, A. (2020). Bob's Guide to Operational Learning: How to Think Like a Human and Organizational Performance (HOP) Coach. Independently published.

Perrow, C. (1999). Normal Accidents: Living with High-Risk Technologies (Revised ed.). Princeton University Press.

Hollnagel, E. (2017). Safety-II in Practice: Developing the Resilience Potentials (1st ed.). Routledge.

Hollnagel, E. (2014). Safety-I and Safety-II: The Past and Future of Safety Management (1st ed.). CRC Press.

Hollnagel, E., Pariès, J., Woods, D., & Wreathall, J. (2013). Resilience Engineering in Practice: A Guidebook (Ashgate Studies in Resilience Engineering) (1st ed.). CRC Press.

Provan, D., & Dekker, S. (2022). A Field Guide to Safety Professional Practice. Safety Futures.

Schein, E. H., & Schein, P. A. (2021). Humble Inquiry, Second Edition: The Gentle Art of Asking Instead of Telling (The Humble Leadership Series) (Expanded ed.). Berrett-Koehler Publishers.

Sutton, B. L., McCarthy, G., Robinson, B. M., Sutton, B., & Conklin, T. (2020). The Practice of Learning Teams: Learning and improving safety, quality, and operational excellence. Independently published.

Edmondson, A. C. (2018). The Fearless Organization: Creating Psychological Safety in the Workplace for Learning, Innovation, and Growth (1st ed.). Wiley.

Schein, E. H., & Schein, P. A. (2016). Organizational Culture and Leadership (The Jossey-Bass Business & Management Series) (5th ed.). Wiley.

And many more...

PODCASTS

Pre-Accident Investigation Podcast
hosted by Todd Conklin

The Safety of Work
hosted by Drew Rae and David Provan

The HOP Nerd
hosted by Sam Goodman

The Jay Allen Show
hosted by Jay Allen

Rebranding Safety
hosted by James MacPherson

Social Psychology of Risk Podcast
hosted by Robert Long

DisasterCast
hosted by Drew Rae

The Practice of Learning Teams
hosted by Brent Sutton et al.

And many more...

WEBSITES

Safety Differently
www.safetydifferently.com

HOP Hub
www.hophub.org

The HOP Nerd
www.thehopnerd.com

Learning Teams Community
www.learningteamscommunity.com

HOP Lab
www.southpacinternational.com/hoplab

Sam Goodman is a father, husband, and a friend. He is also a Human & Organizational Performance practitioner and consultant, safety professional, and betterment evangelist. He is the author of multiple books focused on the safety of work and the safety profession, and the host and producer of The HOP Nerd Podcast and Really F**king Scary Stories. He is the founder of The HOP Nerd which focuses on providing Human & Organizational Performance consulting services and Pale Horse Media Co. Sam is an accomplished author, speaker, consultant, and coach. He lives in Phoenix, Arizona with his husband Jerel and their amazing daughter Avery. Sam enjoys creating "bad ass things" and has made it his life's mission to *"Make the World a Better Place to Work"* by *"making safety suck less."*

BIBLIOGRAPHY

PhD, Conklin Todd. *The 5 Principles of Human Performance: A Contemporary Update of the Building Blocks of Human Performance for the New View of Safety*. Independently published, 2019.

Dekker, Sidney. *Safety Differently: Human Factors for a New Era, Second Edition*. 2nd ed., CRC Press, 2014.

Dekker, Sidney. *Just Culture: Restoring Trust and Accountability in Your Organization, Third Edition*. 3rd ed., CRC Press, 2016.

Bayless, Kate. "What Is Helicopter Parenting?" Parents, 2019, www.parents.com/parenting/better-parenting/what-is-helicopter-parenting.

McCarthy, K., & More, R. (2021, October 29). Are You a Helicopter Parent? Signs and Characteristics to Avoid. LoveToKnow. https://family.lovetoknow.com/parenting-tips-strategies-modern-world/helicopter-parents-facts-characteristics-know

Goodman, Samuel Uriah, and Ian Allison. *Safety Sucks! The Manifesto*. Independently published, 2021.

Goodman, Sam. *WTFRM?: A Reflection on What Is Meaningful to Workplace Safety*. Independently Published, 2021.

Hall, Ph.D., E. D. H. (2019, June 6). Why We Hate People Telling Us What to Do. Psychology Today. Retrieved July 19, 2022, from https://www.psychologytoday.com/us/blog/conscious-communication/201906/why-we-hate-people-telling-us-what-do

Bessarabova, E., Fink, E. L., & Turner, M. (2013). Reactance, restoration, and cognitive structure: Comparative statics. Human Communication Research, 39(3), 339-364.

Dixon. (2016). Compliance: An Introduction. IB PSYCHOLOGY. Retrieved July 19, 2022, from https://www.themantic-education.com/ibpsych/2016/10/25/compliance-an-introduction/

Cherry. (2022, June 8). The Psychology of Compliance. Verywell Mind. https://www.verywellmind.com/what-is-compliance-2795888

Cullum J, O'Grady M, Armeli S, Tennen H. The role of context-specific norms and group size in alcohol consumption

and compliance drinking during natural drinking
events. Basic Appl Soc Psych. 2012;34(4):304-312.
doi:10.1080/01973533.2012.693341

Walton. (2017). Understanding Other People Requires Being
Them, Not Reading Them. The University of Chicago
Booth School of Business. Retrieved July 19, 2022,
from
https://www.chicagobooth.edu/review/understanding-
other-people-requires-being-them-not-reading-them

Haotian Zhou, Elizabeth A. Majka, and Nicholas Epley, "Inferring
Perspective versus Getting Perspective:
Underestimating the Value of Being in Another
Person's Shoes," Psychological Science, February 2017.

Grant. (2015, April 16). We're All Terrible at Understanding Each
Other. Harvard Business Review. Retrieved July 19,
2022, from https://hbr.org/2015/04/were-all-terrible-at-
understanding-each-other

Mcleod, S. (2018). Fundamental Attribution Error. Simply
Psychology.
https://www.simplypsychology.org/fundamental-
attribution.html

Lloyd, C. (2021). Next Generation Safety Leadership (1st ed.).
CRC Press.

Havinga, J.; Shire, M.I.; Rae, A. Should We Cut the Cards?
Assessing the Influence of "Take 5" Pre-Task Risk
Assessments on Safety. Safety 2022, 8, 27.
https://doi.org/ 10.3390/safety8020027

Sutton, B. L., McCarthy, G., Robinson, B., & Conklin, T. (2020).
The Practice of Learning Teams: Learning and
improving safety, quality and operational excellence.
Independently published.

Baker, A. (2022, May 2). A Short Introduction to Human and
Organizational Performance (HOP) and Learning
Teams. Safetydifferently.Com. Retrieved July 20, 2022,
from https://safetydifferently.com/a-short-introduction-
to-human-and-organizational-performance-hop-and-
learning-teams/

Edwards, Bob, et al. "Bob and Andy: HOP Foundation and Intro to
Learning Teams." HOP Hub,
www.hophub.org/_files/ugd/1a0149_4977f9027414499
db5e3e43cc7706a60.pdf. Accessed 21 July 2022.

Edwards, Bob, and Andrea Baker. "Bob and Andy: Learning Team
Deep Dive." HOP Hub,

www.hophub.org/_files/ugd/1a0149_36353f26c2fb4a8
49ccc53a1a21456f4.pdf. Accessed 21 July 2022.

Dekker, Sidney. The Field Guide to Understanding "Human
Error." 3rd ed., CRC Press, 2014.

Munro, E. R. (2015). The Purpose of Pain. Science Features |
Naked Scientists. Retrieved July 21, 2022, from
https://www.thenakedscientists.com/articles/science-
features/purpose-
pain#:%7E:text=It%20provokes%20an%20unconscious
%20physical,of%20nerves%20within%20the%20body.
(2019, December 3). Owww! The science of pain.
Science News Explores. Retrieved July 21, 2022, from
https://www.snexplores.org/article/owww-science-pain

Denial. (n.d.). Psychology Today. Retrieved July 21, 2022, from
https://www.psychologytoday.com/us/basics/denial

Borschel, M. (2021, May 31). Why do people lash out?
Monica Borschel.
https://doctormonicaborschel.com/2019/08/21/why-do-
people-lash-out/

Williams, K. "Chapter 6 Ostracism: A Temporal Need-Threat
Model, Advances in Experimental Social Psychology, Academic
Press, Volume 41, 2009, Pages 275-314,

https://doi.org/10.1016/S0065-2601(08)00406

(https://www.sciencedirect.com/science/article/pii/S006526010800
4061)

Edmondson, A. C. (2018). The Fearless Organization: Creating
Psychological Safety in the Workplace for Learning, Innovation,
and Growth (1st ed.). Wiley.

Psychology Tools. (2022, May 17). Fight Or Flight Response.
https://www.psychologytools.com/resource/fight-or-flight-
response/#:%7E:text=The%20fight%20or%20flight%20response,b
ody%20to%20fight%20or%20flee.

Cosenzo, V. (2021, March 29). When Safety Proves Dangerous.
Farnam Street. Retrieved July 24, 2022, from
https://fs.blog/safety-proves-dangerous/

Risk Compensation. (n.d.). Wikipedia. Retrieved July 24, 2022,
from https://en.wikipedia.org/wiki/Risk_compensation

SafetyRisk Admin. (2017, February 20). Risk Homeostasis
Theory–Why Safety Initiatives Go Wrong. Safety Risk
.Net. Retrieved July 24, 2022, from
https://safetyrisk.net/risk-homeostasis-theorywhy-
safety-initiatives-go-wrong/

Wilde, G.J.S. (2014). Target Risk 3 – Risk Homeostasis in
Everyday Life. Toronto: PDE Publications – Digital
Edition.

Hallowell. (2020). Safety Classification and Learning (SCL)

     Model. Edison Electric Institute.

     https://www.safetyfunction.com/_files/ugd/3b3562_8d8

     8edfefc4d4c8b8c636dc0267a0c42.pdf

Ferro, S. (2016, January 14). The Paradoxical Ways Bike Helmets

     Make Us Less Safe. Mental Floss.

     https://www.mentalfloss.com/article/73670/paradoxical

     -ways-bike-helmets-make-us-less-safe

Gamble T, Walker I. Wearing a Bicycle Helmet Can Increase Risk

     Taking and Sensation Seeking in Adults. Psychological

     Science. 2016;27(2):289-294.

     doi:10.1177/0956797615620784

W. Kip Viscusi, The Lulling Effect: The Impact of Child-Resistant

     Packaging on Aspirin and Analgesic Ingestions, 74

     AEA Papers and Proceedings. 324 (1984) Available at:

     https://scholarship.law.vanderbilt.edu/faculty-

     publications/130

Conklin, T. (2017). Workplace Fatalities: Failure to Predict: A

     New Safety Discussion on Fatality and Serious Event

     Reduction. Independently Published.

Dekker, S. (2018). The Woolworths Experiment. Safety
        Differently. Retrieved July 24, 2022, from
        https://safetydifferently.com/the-woolworths-
        experiment/

Guidelines Work, Rules Don't. (2022). Weidel on Winning.
        https://weidelonwinning.com/blog/guidelines-work-
        rules-dont/

Mishra, T. (2022, April 6). Safety Compliance. Safeopedia.
        https://www.safeopedia.com/definition/3969/safety-
        compliance

OSHA. (n.d.). Commonly Used Statistics | Occupational Safety
        and Health Administration. Retrieved July 25, 2022,
        from https://www.osha.gov/data/commonstats

OSHA. (n.d.-b). Fatality Inspection Data | Occupational Safety and
        Health Administration. Retrieved July 25, 2022, from
        https://www.osha.gov/fatalities

Brown, J. (2020, July 17). Nearly 50 years of occupational safety
        and health data: Beyond the Numbers: U.S. Bureau of
        Labor Statistics. U.S. BUREAU OF LABOR
        STATISTICS. Retrieved July 25, 2022, from
        https://www.bls.gov/opub/btn/volume-9/nearly-50-
        years-of-occupational-safety-and-health-data.htm

Dekker, S. (2019). Foundations of Safety Science: A Century of
Understanding Accidents and Disasters (1st ed.).
Routledge.

OSHA. (n.d.-c). OSHA Penalties | Occupational Safety and Health
Administration. Occupational Safety and Health
Administration. Retrieved July 25, 2022, from
https://www.osha.gov/penalties

NSC. (n.d.). Work Injury Costs. National Safety Council - Injury
Facts. https://injuryfacts.nsc.org/work/costs/work-
injury-costs/

OSHA. (n.d.-c). Fatality Inspection Data | Occupational Safety and
Health Administration. Occupational Safety and Health
Administration. Retrieved July 25, 2022, from
https://www.osha.gov/fatalities

Baker, A. (2019, April 18). An Introduction to the 5 Phases of
HOP Integration. Safety Differently. Retrieved July 26,
2022, from https://safetydifferently.com/an-
introduction-to-the-5-phases-of-hop-integration/

SafetyRisk. (2022, March 5). The Zero Safety Paradox. Safety
Risk.Net. Retrieved July 26, 2022, from
https://safetyrisk.net/the-zero-safety-paradox/

Safeopedia. (2018, September 26). Zero Harm. Safeopedia.Com.

Retrieved July 26, 2022, from

https://www.safeopedia.com/definition/6854/zero-harm

Gerard Zwetsloot, Stavroula Leka & Pete Kines (2017) Vision

zero: from accident prevention to the promotion of

health, safety and well-being at work, Policy and

Practice in Health and Safety, 15:2, 88-100, DOI:

10.1080/14773996.2017.1308701

Fred Sherratt & Andrew R. J. Dainty (2017) UK construction

safety: a zero paradox?, Policy and Practice in Health

and Safety, 15:2, 108-

116, DOI: 10.1080/14773996.2017.1305040

Dekker, Sidney. (2017). Zero commitment: commentary on

Zwetsloot et al., and Sherratt and Dainty. Policy and

Practice in Health and Safety. 15. 1-7.

10.1080/14773996.2017.1374027.

ABC News. (2010, May 5). Louisiana Oil Spill: Feds Gave Safety

Prize to Transocean's Deepwater Horizon.

https://abcnews.go.com/Blotter/louisiana-oil-spill-feds-

gave-safety-prize-transoceans/story?id=10528236

U.S. CHEMICAL SAFETY AND HAZARD INVESTIGATION

BOARD. (2007). INVESTIGATION REPORT -

REFINERY EXPLOSION AND FIRE.

https://www.csb.gov/bp-america-refinery-explosion/

Hallowell, M., Quashne, M., Salas, R., Jones, M., MacLean,B. and Quinn, E. (2020) The statistical invalidity of TRIR as a measure of safety performance. Construction Safety Research Alliance.

Department of Justice. (2013). Former Shaw Group Safety Manager At TVA Nuclear Sites Sentenced To 78. United States Department of Justice. Retrieved July 26, 2022, from https://www.justice.gov/usao-edtn/pr/former-shaw-group-safety-manager-tva-nuclear-sites-sentenced-78-months-prison-major

Conklin, T. (2012). Pre-Accident Investigations: An Introduction to Organizational Safety (1st ed.). CRC Press

Law Insider. (n.d.). Work rule Definition. Retrieved July 26, 2022, from https://www.lawinsider.com/dictionary/work-rule

Staughton, J. (2022, January 22). What Is Malicious Compliance? Science ABC. Retrieved July 26, 2022, from https://www.scienceabc.com/social-science/what-is-malicious-compliance-meaning-examples.html

Usrey, C. (2021, August 29). The Campbell Institute: What are safety leading indicators? Safety+Health. Retrieved July 27, 2022, from

https://www.safetyandhealthmagazine.com/articles/138
21-the-campbell-institute-what-are-safety-leading-
indicators

Eiser, J. R., & Eiser, C. (1975). Prediction of environmental
change: Wish-fulfillment revisited. European Journal of
Social Psychology, 5(3), 315–322.
https://doi.org/10.1002/ejsp.2420050305

Beaton, C. (2017, November 13). Humans Are Bad at Predicting
Futures That Don't Benefit Them. The Atlantic.
Retrieved July 27, 2022, from
https://www.theatlantic.com/science/archive/2017/11/h
umans-are-bad-at-predicting-futures-that-dont-benefit-
them/544709/

Investopedia. (2022, May 26). A Look Into the Great Recession.
Retrieved July 27, 2022, from
https://www.investopedia.com/terms/g/great-
recession.asp#:%7E:text=Key%20Takeaways,Great%2
0Depression%20of%20the%201930s.

Vector Solutions. (2021, June 12). Preventing Workplace Fatalities
(Based on Dr. Todd Conklin's Book "Workplace
Fatalities: Failure to Predict").
https://www.vectorsolutions.com/resources/blogs/preve
nting-workplace-fatalities/

350

Lehrer, E. (2019). America has too many criminal laws. The Hill.
Retrieved July 27, 2022, from
https://thehill.com/opinion/criminal-justice/473659-
america-has-too-many-criminal-laws/

Baker, A. (2019, April 18). An Introduction to the 5 Phases of
HOP Integration. Safety Differently. Retrieved July 28,
2022, from https://safetydifferently.com/an-
introduction-to-the-5-phases-of-hop-integration/

Mitchell, M. (2022, February 25). High Performance Or
Humanity? Leaders Must Embrace Both. Forbes.
Retrieved July 28, 2022, from
https://www.forbes.com/sites/forbescoachescouncil/202
2/02/23/high-performance-or-humanity-leaders-must-
embrace-both/?sh=646fa88238ec

Besnard, Denis & Hollnagel, Erik. (2012). Some myths about
industrial safety.

Hendricks, D., Fell, J., Freedman, M. (2001) The Relative
Frequency of Unsafe Driving Acts in Serious Traffic
Crashes [Summary Report] Published Date : 2001-01-
01 Report Number : DOT-HS-809-205;NTIS-
PB2001104249; DOI :
https://doi.org/10.21949/1525533

Statista. (2021, August 4). Fatality rate per 100,000 drivers

 licensed in the U.S. 1990–2019. Retrieved July 29,

 2022, from

 https://www.statista.com/statistics/191660/fatality-rate-

 per-100000-licensed-drivers-in-the-us-since-1988/

Woods, D. D., Johannesen, L. J., Cook, R. I. & Sarter, N. B.

 (1994). Behind human error: Cognitive systems,

 computers and hindsight. Columbus, OH: CSERIAC

INDEX

Justice, 108, 138

positive, 74
micro-experiment, 53
minimization, 98
misanthropic, 42
missteps, 34
human, 118
modification, 51–52
monitoring people, 38
monocausality, 60
myths, 88, 139

N
negative consequences, unintended, 36
negative effects, 79
neutral position, 43
non-brittle, 83
non-compliance, 33, 87–88, 99, 102
norms, 89, 133

O
observation, increased, 14
observation cards, 15
documented behavioral safety, 77
obsession, 82
Occupational Safety, 136
Occupational Safety and Health Administration. *See* OSHA
occurrences, preventing lower outcome, 112
operational conditions, 89
operational intelligence, vital, 27, 57, 106
operational learning, 55, 62, 120, 124, 126, 131
operational learning principles, 56
operational surprises, 39, 56, 118, 124
unintended, 120
operators, 74, 83
organizational approaches, 43
common, 107

symptoms, 48, 71, 73, 102
fixing surface-level, 22
systems, organizational, 118

T

tactics, 18, 24, 33, 117, 121–22
brutal enforcement, 33
teams, 51–52, 58, 104
Texas City Refinery explosion, 106
tooling, 74
tools
effective, 92–93
vital safety, 50
total recordable incident rate (TRIR), 106, 138
traditional approaches, 8, 13, 15, 18–19, 24, 26, 86, 88,
    97–98, 101, 111–12
traditional counterpoints, 26
Traditional definitions, 104
traditional ideas, 18, 25
traditional safety, 13, 15–16, 25, 116
traditional safety organizations, 4
trust, 9, 26, 32, 37–39, 42–43
trusting, 26, 42–43
Typical organizational approaches, 107

U

understanding, 3, 7, 39, 41, 46–48, 56, 60, 89, 94, 131,
    134
unintended consequences, 37, 47, 79–82
untrusting, 8, 34, 118
user
intended, 92
primary, 51

V

vertical command, 9
violation, 40, 87, 95

Printed in Great Britain
by Amazon